Lecture Notes in Mathematics

Volume 2379

This series reports on new developments in all areas of mathematics and their applications - quickly, informally and at a high level. Mathematical texts analysing new developments in modelling and numerical simulation are welcome. The type of material considered for publication includes:

1. Research monographs
2. Lectures on a new field or presentations of a new angle in a classical field
3. Summer schools and intensive courses on topics of current research.

Texts which are out of print but still in demand may also be considered if they fall within these categories. The timeliness of a manuscript is sometimes more important than its form, which may be preliminary or tentative. Please visit the LNM Editorial Policy (https://drive.google.com/file/d/1MOg4TbwOSok RnFJ3ZR3ciEeKs9hOnNX_/view?usp=sharing)

Titles from this series are indexed by Scopus, Web of Science, Mathematical Reviews, and zbMATH.

Javier Alejandro Chávez-Domínguez •
Verónica Dimant • Daniel Galicer

Operator Space Tensor Norms

 Springer

Javier Alejandro Chávez-Domínguez
David and Judi Proctor Department of
Mathematics
University of Oklahoma
Norman, OK, USA

Verónica Dimant
Departamento de Matemática y Ciencias
Universidad de San Andrés
Victoria, Buenos Aires, Argentina

Daniel Galicer
Departamento de Matemáticas y Estadística
Universidad Torcuato Di Tella
Buenos Aires, Argentina

ISSN 0075-8434   ISSN 1617-9692   (electronic)
Lecture Notes in Mathematics
ISBN 978-3-031-96208-0   ISBN 978-3-031-96209-7   (eBook)
https://doi.org/10.1007/978-3-031-96209-7

Mathematics Subject Classification: 47L25, 46M05, 46B28

This work was supported by National Science Foundation (DMS-1900985 and DMS-2247374 Alejandro Chavez Dominguez), Consejo Nacional de Investigaciones Científicas y Técnicas (11220200101609CO Verónica Dimant, 11220200102366CO Daniel Galicer) and Agencia Nacional de Promoción Científica y Tecnológica (PICT 2018-4250 Daniel Galicer, PICT 2018-04104 Verónica Dimant).

This Springer imprint is published by the registered company Springer Nature Switzerland AG
The registered company address is: Gewerbestrasse 11, 6330 Cham, Switzerland

If disposing of this product, please recycle the paper.

Declarations

Competing Interests The authors have no competing interests to declare that are relevant to the content of this manuscript.

Introduction

In the 1930s von Neumann in collaboration with Murray began a "mathematical quantization program," aiming to formulate an operator version of integration theory. They were successful, with the key idea being to replace function spaces by $*$-algebras of bounded operators on Hilbert spaces. This was the seed that started a long road of "noncommutative analysis" which lead (among many other things) to what is now known as the theory of operator spaces. Along this road and prior to the formalization of this theory, many concepts which are now typical in this framework (e.g., completely bounded mappings) were introduced in the context of C^*-algebras.

A (Banach) operator space is a Banach space E with an extra "matricial norm structure": in addition to the usual norm on E, we have norms on all the spaces $M_n(E)$ of $n \times n$ matrices with entries from E (often called the matrix levels), where these norms must satisfy certain consistency requirements. The connection with operator algebras comes from the fact that when E is a subspace of an algebra $\mathcal{B}(H)$ of bounded linear operators on a Hilbert space H, there is a canonical norm on $M_n(E)$ coming from the identification between $M_n(\mathcal{B}(H))$ and $\mathcal{B}(H^n)$. The natural morphisms in this category are no longer the maps which are just bounded, but rather the *completely bounded ones*: they are required to be uniformly bounded on all the matrix levels. Operator spaces are thus a quantized or noncommutative version of Banach spaces, giving rise to a theory that not only is mathematically attractive but it is also naturally well-positioned to have applications to quantum physics. The systematic study of operator spaces begun with Ruan's work, and was deepened mainly by Effros and Ruan, Blecher and Paulsen, Pisier and Junge (see the monographs [12, 34, 44, 64, 65] and the references therein). Operator spaces have often provided an appropriate framework to develop many areas of noncommutative analysis.

It is quite natural to investigate to what extent the classical theory of Banach spaces can be translated to the noncommutative context, and this has been extensively studied throughout the years. Though some properties do carry over, many do not and these differences are one of the reasons making the new theory so interesting and rich. And even for those properties that do admit a generalization, the proofs are often different and require the development of new tools. Along these lines, the main goal of this monograph is to take further steps toward the systematic development of a complete theory of tensor products and tensor norms in the operator space setting, and its connections with the parallel theory of their associated mapping ideals.

It should be noted that both ideals of linear operators and tensor norms have inspired important developments in the operator space setting. Noncommutative versions of nuclear, integral, summing, and other mapping ideals have played significant roles in various works, such as [34, 37, 44, 45, 64]. Similarly, several tensor norms for operator spaces, most notably the Haagerup norm, have also been influential; for a good summary, see [12, Sec. 1.5].

Blecher and Paulsen [13] have showed that part of elementary and basic theory of tensor norms of Banach spaces carries over to operator spaces, initiating a "tensor norm program" for operator spaces further developed in [10]. However, despite the depth and elegance of the abstract framework developed in the monograph by Defant and Floret [23] on operator ideals and tensor norms, many of its tools do not appear to have been treated yet in the operator space context. Moreover, while it is clear to specialists that there are obstacles making it impossible to fully translate the classical tensor product theory into the operator space setting, it has not been highlighted exactly which portions do translate and which ones do not. Hence, we will focus on seeing to what extent the classical theory can be transferred. In many situations, this is not the case and additional assumptions about the structure of the operator spaces or the tensor norms involved are really necessary. A good example is local reflexivity: since all Banach spaces have this property but not all operator spaces do, it is well known that operator space versions of classical results will often need local reflexivity as an additional assumption. However, when it comes to tensor norms it turns out that in some cases, we can still get operator space versions of some classical results without requiring local reflexivity, as long as the tensor norm satisfies some extra condition.

Although the theory of operator spaces has its own peculiarities, the classical theory will be our reference and roadmap to obtain and develop noncommutative versions of the theory of tensor norms. Therefore, let us now briefly recall the history of the development of classical theory. Tensor products burst into the world of functional analysis in the late 1930s with the works of Murray, von Neumann, and Schatten, but it was Grothendieck who first truly revealed the richness of properties for tensor products. His famous article *"Résumé de la théorie métrique des produits tensoriels topologiques"* [40] established the basis of what we know today as "local theory"—that is, the study of Banach spaces in terms of their finite-dimensional subspaces—and showed the relevance of the use of tensor products in the theory of normed spaces. All of this helped him to establish a very fruitful theory of duality.

While nowadays the *"Résumé"* is considered foundational work in the area, this masterpiece remained mostly unnoticed for many years. It was not until the end of the 1960s when it started to be better valued by the scientific community. Lindenstrauss and Pełczyński [49] presented some important connections and applications to the theory of absolutely p-summing operators, translating results written in terms of certain tensor norms by Grothendieck into properties of operator ideals. On the other hand, almost simultaneously, a general theory of operator ideals on the class of Banach spaces was developed by Pietsch (and his school in Jena) but without the use—and the language—of tensor norms. Pietsch's book *"Operator Ideals"* [58], which appeared at the end of the seventies, led this theory to become one of the central themes of study in Banach space theory.

At that time, functional analysts generally preferred the language of operator ideals to that of tensor products, so the former received more attention. The theory of tensor products in Banach spaces became stronger and more ubiquitous during the 1980s with the seminal work of Pisier [61, 62], which showed the usefulness of these tools.

With time, the theory of tensor norms became an interesting field in its own right. Defant and Floret, in their famous monograph *"Tensor Norms and Operator Ideals"* [23], made it clear that the theory of tensor products and the theory of operator ideals are two sides of the same coin. Their work had a tremendous legacy: nowadays authors use indistinctly both languages. This same approach has also been developed in various other books on tensor products of Banach spaces [27, 71].

As stated in [23], "both theories, the theory of tensor norms and of normed operator ideals, are more easily understood and also richer if one works with both simultaneously." With this in mind, certain problems are more easily handled using either the *categorical perspective* due to Pietsch, or Grothendieck's *cycle of ideas on tensor products*. With the present work, it is our intent to bring the benefits of the aforementioned parallel perspectives to the operator space setting. As mentioned above both the approaches of ideals and tensor norms have been repeatedly used in the operator space literature, but there has not been a methodical exploration of the connections between the two. It is worth mentioning that this work does not aim to be self-contained, nor does it pretend to be an extensive monograph like [23]. Some of the theory related to this area is either briefly mentioned or cited without providing full proofs. Our purpose is more modest: to initiate a program that, over time, can be enriched with the contributions of the operator space community. We focus on highlighting key results and insights, particularly those relevant to the systematic development of tensor products and tensor norms in the operator space setting.

Thus, numerous topics covered in [23] have been left out. Some expected results that do not appear in the literature can be found in this monograph, and also some new insights that do not have a counterpart or differ from the theory of Banach spaces.

The manuscript is structured as follows. Chapter 1 provides the necessary background material. We introduce the notation, present the basic definitions used throughout the manuscript, and review key aspects of the classical operator space tensor norms—specifically, the minimal, projective, and Haagerup norms. We also discuss vector-valued Schatten spaces, including their representation as Haagerup tensor products and various norm computations for matrices of their elements. Additionally, we emphasize the main relevant differences between the operator space and the Banach space frameworks, which have an impact in the study of tensor norms in both settings.

In Chap. 2, we introduce our main object of study: operator space tensor norms (o.s. tensor norms for short). We begin by establishing some fundamental properties and reviewing existing examples from the literature. We then provide a more precise description of the Chevet-Saphar o.s. tensor norms studied in [18]. Further, we discuss λ-tensor products, developed by Defant and Wiesner in [24, 75], and introduce a new family of Haagerup-style tensor norms. Finally, we explore two useful constructions: the sum and intersection of o.s. tensor norms.

In Chap. 3, we focus on the concepts of finitely and cofinitely generated o.s. tensor norms. A tensor norm is finitely generated if it can be approximated using finite-dimensional subspaces, whereas it is cofinitely generated if such approximations can be obtained through quotients by subspaces of finite codimension. We also define the finite and cofinite hull of a given o.s. tensor norm. Furthermore, we provide several examples and establish that the property of being finitely generated is preserved under intersections and sums.

Chapter 4 deals with the "Five Basic Lemmas" (see Section 13 in [23]) for the operator space setting. These are the *Approximation Lemma*, the *Extension Lemma*, the *Embedding Lemma*, the *Density Lemma*, and the $\mathcal{L}_p$-*Technique Lemma*. To prove some of these results, unlike the classical case, we additionally need certain strong hypotheses such as local reflexivity. However, in the case of λ-o.s. tensor norms this extra assumption is not necessary. Moreover, we introduce a larger class of *extended λ-tensor norms*, for which the lemmas hold without the local reflexivity condition.

In Chap. 5, we introduce dual tensor norms and establish the duality between the intersection and sum procedures. We develop an operator space version of the *Duality Theorem*, which, roughly speaking, relates the dual of a tensor product to the tensor product of the dual spaces. As expected, this result once again requires a local reflexivity assumption, except for the extended λ-tensor norms. We present right- and left-accessible tensor norms and show that every λ-o.s. tensor norm is accessible. Furthermore, to establish certain duality relations, we define the concept of local accessibility—a weaker form of accessibility that applies to locally reflexive operator spaces. Also, we give noncommutative versions of the Chevet-Persson-Saphar inequalities. These inequalities relate the tensor product $S_p(H)\otimes E$, equipped with Chevet–Saphar o.s. tensor norms to the vector-valued Schatten class $S_p[H; E]$. Finally, we conclude the chapter with Sect. 5.5 which reveals connections between our o.s. tensor norms and analysis on the quantum Boolean cube, specifically hypercontractive inequalities. Beyond the intrinsic interest of the results presented here—which also have direct connections to Quantum Information Theory—this section demonstrates how the theoretical framework developed in this manuscript can potentially be useful in other contexts. Ultimately, the tensor-based perspective provides a powerful and versatile tool, offering new avenues for research and applications.

In Chap. 6 we discuss completely metric and completely bounded approximation properties and relate these properties with tensor products, in particular the weak density of a multiple of $B_{E'\otimes_{\min}F}$ in $B_{\mathrm{CB}(E,F)}$. Also, we consider a weak* completely bounded approximation property (generalizing the metric version studied by

Effros and Ruan [31]) which is characterized by having a "nice" duality between the projective and the minimal o.s. tensor norms.

The notion of mapping ideal is the focus of Chap. 7, where we present an extensive list of examples already studied in the literature. To maintain the core of the Banach space concept and allow the natural relationship with tensor norms, we have to slightly strengthen the existing definition of mapping ideal from [34, Sec. 12.2] and add an extra condition. Also, we introduce some typical procedures: injective/surjective hulls of a given mapping ideal, the dual mapping ideal, and present some properties.

Chapters 8 and 9 deal with maximal and minimal mapping ideals and their *Representation theorems*. As the names indicate, maximal and minimal ideals associated to a given tensor norm are the largest and smallest ideals (with respect to the inclusion) which agree with the tensor norm over finite dimensional spaces. The representation theorem for maximal mapping ideals shows how the ideal can be seen as a dual of a tensor product. Again, the hypothesis of local reflexivity is needed here in order to obtain a full characterization. As a consequence, we prove an *Embedding Theorem* (which gives a natural completely isometric inclusion from a tensor product into the mapping ideal) and show how dual and adjoint ideals behave in terms of their associated tensor norms. For minimal ideals, the representation theorem provides a natural complete quotient mapping from a tensor product onto the mapping ideal. Therefore, in a sense, both theorems (for maximal and minimal ideals) relate a mapping ideal with a tensor norm associated to it.

In Chap. 10, we introduce the definitions of right and left completely injective and completely projective hulls of an o.s. tensor norm. Roughly speaking, these procedures are related to preserving complete isometries and complete quotients, respectively. We explore some of their key properties and investigate conditions under which they are finitely generated. Additionally, we define (bilateral) completely injective and completely projective hulls, proving that in both cases, the left and right procedures commute. We explore the possibility of providing a local description of the projective and injective hulls of an o.s. tensor norm. For the first one, this is indeed possible; however, we show that one cannot expect an analogous result, as in the classical setting, for the injective hull. However, this can be achieved by adding certain hypotheses to the space.

We establish in Chap. 11 that, unlike what happens in the Banach space setting, accessibility is a required property of a tensor norm in order to obtain the usual relations between duality and injective/projective hulls. Accessibility is also needed to associate a left (or right) injective hull of a norm to the surjective (respectively, injective) hull of a mapping ideal. Based on the classical theory, we also provide a definition of accessibility for mapping ideals. As expected, right-accessible finitely generated tensor norms are associated with right-accessible mapping ideals but, surprisingly, the left version of this result does not hold. We see that a left-accessible o.s. tensor norm might have an associated mapping ideal which is not left-accessible (although we show in Proposition 11.3.4 that, in this case, the mapping ideal does have a weaker version of left-accessibility).

Chapter 12 deals with the structure of *natural* norms, in the sense of Grothendieck. We prove that some results from the Banach space theory about equivalences or inequalities between injective and projective hulls remain valid in this new context but the whole picture is undoubtedly substantially different. In the classical theory, Grothendieck's inequality (which for instance implies that there is not possible for a tensor norm to be both injective and projective) has a strong impact in the description of natural tensor norms. The existence of an injective and projective Haagerup tensor norm shows us that the situation in the operator space setting could not be at all a translation from the classical case. We use the term proj *family* for the set of norms obtained from the norm proj after taking left or right injective/projective hulls finitely many times. Analogously, we define the min *family*. Unlike the classical setting, these families are not intertwined: each member of the proj family dominates any member of the min family. We present incomplete pictures of both of these families, proving some dominations and non-equivalences and leaving a large list of open questions about them. On the other hand, we completely describe the list of all natural norms that come from applying to min or proj two-sided hull operations (injective or projective). Precisely, it consists of six o.s. tensor norms. Again, this differs from the Banach space case where there are actually four (see Sect. 12.4).

We finish this monograph with Chap. 13, which provides some conclusions of the present work and proposes several open questions for future research.

Contents

Chapter 1
Preliminaries

In this chapter, we present some fundamental concepts of the theory of operator spaces. While we assume familiarity with the basics, we recall certain elementary definitions for completeness. Excellent references on the topic include [12, 34, 54, 65]. The book [67] also deals with tensor products of operator spaces, although concentrating on the case of C^*-algebras. Our notation follows closely that from [64, 65]. For technical reasons, unlike the general literature in the subject, we will not assume that our operator spaces are complete. Thus, to avoid confusion, when they are complete, we choose to call them *Banach operator spaces*. On the other hand, when completeness is not assumed, we refer to them as *normed operator spaces* or, sometimes, for simplicity, just *operator spaces*. The letters E, F and G will always denote normed operator spaces, that is, normed vector spaces with an additional structure at the matricial level.

1.1 The Basics

As usual, $M_{n,m}(E)$ will stand for the set of $n \times m$-matrices of elements in E. In the case $n = m$, we abbreviate it to $M_n(E)$. For $a \in M_{n,m}$ ($n \times m$-matrices of complex scalars), $\|a\|$ denotes its norm as an operator from ℓ_2^m to ℓ_2^n, where, as usual, ℓ_2^k represents the k-dimensional Hilbert space.

Given $x = (x_{ij}) \in M_n(E)$ and $y = (y_{kl}) \in M_m(E)$, we consider $x \oplus y \in M_{n+m}(E)$ to be the matrix $x \oplus y = \begin{pmatrix} (x_{ij}) & 0 \\ 0 & (y_{kl}) \end{pmatrix}$.

Definition 1.1.1 E is a *normed operator space* if, for each n, there is a norm $\| \cdot \|_n$ on $M_n(E)$ satisfying Ruan's Axioms:

M1 $\|x \oplus y\|_{M_{n+m}(E)} = \max \left\{ \|x\|_{M_n(E)}, \|y\|_{M_m(E)} \right\}$, for all $x \in M_n(E)$ and $y \in M_m(E)$.

© The Author(s), under exclusive license to Springer Nature Switzerland AG 2025
J. A. Chávez-Domínguez et al., *Operator Space Tensor Norms*, Lecture Notes in Mathematics 2379, https://doi.org/10.1007/978-3-031-96209-7_1

M2 $\|axb\|_{M_n(E)} \leq \|a\| \cdot \|x\|_{M_m(E)} \cdot \|b\|$, for all $x \in M_m(E)$, $a \in M_{n,m}$ and $b \in M_{m,n}$.

If H is a Hilbert space and $E \subset \mathcal{B}(H)$ is a subspace, there is a natural norm in $M_n(E)$ given through the identification $M_n(\mathcal{B}(H)) = \mathcal{B}(H^n)$, which endows E with a normed operator space structure. On the other hand, by Ruan's representation theorem [34, Thm. 2.3.5] for any operator space E there exist a Hilbert space H and an inclusion $E \subset \mathcal{B}(H)$ which is an isometry for all the matrix levels.

We denote by ONORM and OBAN the classes of all normed and Banach operator spaces, respectively. Similarly, we write OFIN for the class of all finite-dimensional operator spaces.

Every linear mapping $T : E \to F$ induces, for each $n, m \in \mathbb{N}$, a linear mapping $T_{n,m} : M_{n,m}(E) \to M_{n,m}(F)$ (called the *amplification of T*) given by

$$T_{n,m}(x) = \left(T(x_{ij})\right), \text{ for all } x = (x_{ij}) \in M_{n,m}(E).$$

Once again, in the case $n = m$ we write T_n instead of $T_{n,n}$.

The *completely bounded norm* of T is defined by

$$\|T\|_{\mathrm{cb}} = \sup_{n \in \mathbb{N}} \|T_n\| = \sup_{n \in \mathbb{N}} \left\{ \|T_n(x)\|_{M_n(F)} : \|x\|_{M_n(E)} \leq 1 \right\}.$$

A mapping T is called *completely bounded* when $\|T\|_{\mathrm{cb}}$ is finite and we denote by $\mathrm{CB}(E, F)$ the space of completely bounded mappings from E to F. This is an operator space with the structure given by the identification $M_n(\mathrm{CB}(E, F)) = \mathrm{CB}(E, M_n(F))$. In other words, each $n \times n$ matrix of mappings from E into F can be identified with a mapping from E into the space of $n \times n$ matrices with coefficients in F. Thus, to compute its norm, we apply a calculation as above.

We say that T is a *complete isometry* if each $T_n : M_n(E) \to M_n(F)$ is an isometry and we write $E = F$ to indicate that E and F are completely isometrically isomorphic. A completely isometric embedding $T : E \to F$ will be called a *complete injection* for short.

The mapping $T : E \to F$ is said to be a *complete quotient* or *complete metric surjection* or *complete projection* if it is onto and the associated map from $E/\ker(T)$ to F is a completely isometric isomorphism. In [65, Sec. 2.4], it is proved that a linear map $T : E \to F$ is a complete quotient if and only if its adjoint $T' : F' \to E'$ is a completely isometric embedding. Note that if a linear map $T : E \to F$ between normed operator spaces is a complete quotient, then for every $x \in M_n(F)$ we have

$$\|x\|_{M_n(F)} = \inf \left\{ \|y\|_{M_n(E)} : y \in M_n(E), T_n(y) = x \right\}.$$

The operator space structure of a dual space E' comes from the identification $M_n(E') = \mathrm{CB}(E, M_n)$. The norm of a matrix in $M_n(E')$ can also be computed through the *matrix pairing* $\langle\langle \cdot, \cdot \rangle\rangle : M_n(E) \times M_n(E') \to M_{n^2}$, where $\langle\langle (x_{ij}), (\varphi_{kl}) \rangle\rangle = (\varphi_{kl}(x_{ij}))$. Precisely,

$$\|(\varphi_{kl})\|_{M_n(E')} = \sup \left\{ \|\langle\langle (x_{ij}), (\varphi_{kl}) \rangle\rangle\|_{M_{n^2}} : \|(x_{ij})\|_{M_n(E)} \leq 1 \right\}.$$

Given a bilinear mapping $\phi : E \times F \to G$, we denote by $\phi_n : M_n(E) \times M_n(F) \to M_{n^2}(G)$ the mapping defined, for each $n \in \mathbb{N}$, as follows:

$$\phi_n(v, w) = \left(\phi(v_{ij}, w_{kl})\right), \quad \text{for all } v = (v_{ij}) \in M_n(E), \, w = (w_{kl}) \in M_n(F).$$

We say that ϕ is *jointly completely bounded* when

$$\|\phi\|_{\mathrm{jcb}} \equiv \sup_{n \in \mathbb{N}} \|\phi_n\| < \infty. \tag{1.1.1}$$

The space $\mathrm{JCB}(E \times F, G)$ of all jointly completely bounded bilinear mappings from $E \times F$ to G has an operator space structure given by the identification

$$M_n\left(\mathrm{JCB}(E \times F, G)\right) = \mathrm{JCB}\left(E \times F, M_n(G)\right).$$

Another natural associate to ϕ is the bilinear mapping $\phi_{(n)} : M_n(E) \times M_n(F) \to M_n(G)$ related with the matrix product and given by

$$\phi_{(n)}(v, w) = \left(\sum_{k=1}^{n} \phi(v_{ik}, w_{kl})\right), \quad \text{for all } v = (v_{ij}) \in M_n(E), \, w = (w_{kl}) \in M_n(F).$$

The bilinear mapping ϕ is *multiplicatively bounded* if

$$\|\phi\|_{\mathrm{mb}} = \sup_{n \in \mathbb{N}} \|\phi_{(n)}\| < \infty. \tag{1.1.2}$$

Again, for the space $\mathrm{MB}(E \times F, G)$ of all multiplicatively bounded bilinear mappings from $E \times F$ to G an operator space structure is provided via the identification

$$M_n\left(\mathrm{MB}(E \times F, G)\right) = \mathrm{MB}\left(E \times F, M_n(G)\right).$$

1.1.1   Some Usual Notation

For a normed operator space E, we denote by $\mathrm{OFIN}(E)$ (resp. $\mathrm{OCOFIN}(E)$) the set of all finite-dimensional (resp. finite-codimensional) subspaces of E. Given $L \in \mathrm{OCOFIN}(E)$, let $q_L^E : E \to E/L$ be the canonical projection and given a subspace F of E let $i_F^E : F \to E$ be the canonical injection. To avoid an overload of notation, when the inclusion is the typical complete isometry for operator spaces $E \subset \mathcal{B}(H)$ we just denote the canonical injection by $i_E : E \to \mathcal{B}(H)$. Also, following the usual Banach space notation, for a normed operator space E, $\kappa_E : E \to E''$ denotes the canonical injection into the bidual.

Given vector spaces E and F, we identify a linear map $T : E \to F'$ with a bilinear map $\beta_T : E \times F \to \mathbb{C}$ via $\beta_T(x, y) = (Tx)(y)$; note that β_T can also be identified with an element of $(E \otimes F)'$.

Recall that for a Hilbert space H the Schatten class $S_p(H)$ is defined for $1 \le p < \infty$ as the space of all compact operators T on H such that $\mathrm{tr}\left(|T|^p\right) < \infty$ equipped with the norm $\|T\|_{S_p} = \left(\mathrm{tr}\left(|T|^p\right)\right)^{1/p}$; in the case $p = \infty$, we denote by $S_\infty(H)$ the space of all compact operators on H (also denoted by $\mathcal{K}(H)$) endowed with the operator norm. To provide the spaces $S_p(H)$ with an operator space structure, we follow Pisier's approach [64]: since $S_\infty(H)$ is a C^*-algebra it has a canonical operator space structure [34, p. 21], by the duality $S_1(H)' = \mathcal{B}(H)$ we also get a natural operator space structure on $S_1(H)$, and by complex interpolation for operator spaces [65, Sec. 2.7] we get an operator space structure for each of the intermediate $S_p(H)$ spaces. For simplicity we write S_p instead of $S_p(\ell_2)$, and S_p^n instead of $S_p(\ell_2^n)$.

A normed operator space E is an $\mathscr{OS}_{p,C}$ space [46, Sec. 2] if there is a family $(F_i)_{i \in I}$ of finite-dimensional subspaces of E whose union is dense in E and such that for every index i there is a natural number n_i such that $d_{\mathrm{cb}}(S_p^{n_i}, F_i) \le C$ where d_{cb} stands for the completely bounded Banach-Mazur distance , i.e.,

$$d_{\mathrm{cb}}(E, F) := \inf\left\{\|T\|_{\mathrm{cb}}\|T^{-1}\|_{\mathrm{cb}} : T \in \mathrm{CB}(E, F) \text{ is a complete isomorphism}\right\}.$$

If E is an $\mathscr{OS}_{p,C'}$ space for every $C' > C$, we say that E is an $\mathscr{OS}_{p,C+}$ space .

If W is a Banach space, its *maximal operator space structure* $\mathrm{Max}(W)$ is defined by, for $A \in M_n(W)$

$$\|A\|_{M_n(\mathrm{Max}(W))} = \left\{ \|u_n(A)\|_{M_n(\mathcal{B}(H))} : \|u : W \to \mathcal{B}(H)\| \le 1\right\}.$$

See [65, Chap. 3] for more details.

1.1.2 Direct Sums

We will now recall some necessary facts on direct sums of operator spaces from [65, Sec. 2.6].

Given a family $\{E_\gamma\}_{\gamma \in \Gamma}$ of operator spaces, its ℓ_∞ direct sum which we will denote by $\ell_\infty(\{E_\gamma : \gamma \in \Gamma\})$ is the direct sum $E = \bigoplus_{\gamma \in \Gamma} E_\gamma$ with the operator space structure given as follows: for $u = (u_\gamma)_{\gamma \in \Gamma} \in M_n(E)$,

$$\|u\|_{\ell_\infty(\{E_\gamma \,:\, \gamma \in \Gamma\})} = \sup\left\{\|u_\gamma\|_{M_n(E_\gamma)} : \gamma \in \Gamma\right\}.$$

When we have only two summands, we will often use the notation $E_1 \oplus_\infty E_2$ instead. We denote by $c_0(\{E_\gamma : \gamma \in \Gamma\})$ the subspace of $\ell_\infty(\{E_\gamma : \gamma \in \Gamma\})$ of elements $u = (u_\gamma)_{\gamma \in \Gamma}$ such that the net $(\|u_\gamma\|)_{\gamma \in \Gamma}$ converges to 0 (with respect to the directed set of the finite subsets of Γ).

The ℓ_1-sum of $\{E_\gamma\}_{\gamma \in \Gamma}$ will be similarly denoted by $\ell_1(\{E_\gamma : \gamma \in \Gamma\})$; for simplicity, below we only give the full details in the case $E_1 \oplus_1 E_2$ where we have two summands. For $u \in M_n(E_1 \oplus E_2)$,

$$\|u\|_{M_n(E_1 \oplus_1 E_2)} = \sup \left\{ \|(T_1 \oplus T_2)_n(u)\|_{M_n(\mathcal{B}(H))} : \|T_j : E_j \to \mathcal{B}(H)\|_{\mathrm{cb}} \leq 1 \right.$$
$$\left. \text{for } j = 1, 2 \right\}$$

where $T_1 \oplus T_2 : E_1 \oplus E_2 \to \mathcal{B}(H)$ is the mapping $(T_1 \oplus T_2)(x, y) = T_1(x) + T_2(y)$.

The explicit description above of the operator space structure of an ℓ_1-sum of operator spaces $\ell_1(\{E_\gamma\}_{\gamma \in \Gamma})$ is a little unwieldy. Oftentimes, it is easier to use that this operator space structure is characterized by the following universal property: for any operator space F and any linear map $T : \ell_1(\{E_\gamma\}_{\gamma \in \Gamma}) \to F$, we have $\|T\|_{\mathrm{cb}} \leq 1$ if and only if for all $\gamma \in \Gamma$ we have $\|T J_\gamma\|_{\mathrm{cb}} \leq 1$, where $J_\gamma : E_\gamma \to \ell_1(\{E_\gamma\}_{\gamma \in \Gamma})$ is the canonical injection (that is, the map sending $v \in E_\gamma$ to the vector having v in the γ-th position and 0 everywhere else). Note that as an immediate consequence of the aforementioned universal property, the canonical projections $\ell_1(\{E_\gamma\}_{\gamma \in \Gamma}) \to E_{\gamma_0}$ are completely contractive, and therefore the canonical injections are complete isometries.

1.1.3 Injections and Projections

A normed operator space E is said to be *completely injective* if whenever $F \subseteq G$ are operator spaces and $T : F \to E$ is a completely bounded linear map, there exists an extension $\widetilde{T} : G \to E$ with $\|T\|_{\mathrm{cb}} = \|\widetilde{T}\|_{\mathrm{cb}}$. By the Arveson extension theorem [34, Thm. 4.1.5], $\mathcal{B}(H)$ is completely injective for any Hilbert space H. Let us now observe that any completely injective normed operator space E has to be complete. Indeed, consider the diagram

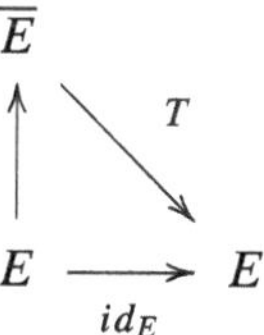

where the vertical arrow is simply the inclusion of E in its completion $\overline{E}$, and the operator T is given by the injectivity of E. If we take a Cauchy sequence $(v_n)_n$ in E, and we take its limit $v \in \overline{E}$, then $T(v) \in E$ is the limit of $(v_n)_n$ since

$$T(v) = T\left(\lim_{n \to \infty} v_n\right) = \lim_{n \to \infty} T(v_n) = \lim_{n \to \infty} v_n.$$

A normed operator space E is said to be *completely projective* if for any completely bounded linear map $T : E \to G/F$ into a quotient space and any

$\varepsilon > 0$ there exists a lifting $\widetilde{T} : E \to G$ of T with $\|\widetilde{T}\|_{cb} \leq (1+\varepsilon)\|T\|_{cb}$. The basic examples of completely projective operator spaces are the spaces S_1^n for any $n \in \mathbb{N}$, and we will also use the fact that an ℓ_1-sum of completely projective operator spaces is again completely projective.

Every Banach operator space E can be seen as a quotient of a completely projective space. Indeed, there is a set I and a family $(n_i)_{i \in I} \subset \mathbb{N}$ such that E is the quotient of $\ell_1(\{S_1^{n_i} : i \in I\})$ (see e.g., [65, Prop. 2.12.2]). We denote the latter space by Z_E, the completely projective cover of E, and $q_E : Z_E \twoheadrightarrow E$ the corresponding complete quotient mapping. The fact that Z_E is completely projective can be tracked in [65, Chap. 24]. We also refer to [65, Chap. 24] for further details on completely injective and completely projective operator spaces.

1.2 Classical Operator Space Tensor Norms

We begin by recalling the definitions of the three most common tensor norms in operator space theory: the minimal, the projective, and the Haagerup tensor norms. Each of these procedures endows the algebraic tensor product $E \otimes F$ with an operator space structure, leading to the operator spaces $E \otimes_{\min} F$, $E \otimes_{\text{proj}} F$, and $E \otimes_h F$, respectively. We enlist their basic properties which can be found in the classical books by Effros and Ruan [34] or Pisier [67]. While these authors always work within the setting of Banach operator spaces, the properties we list below hold even without assuming completeness. As usual, the completions of the three mentioned tensor products are denoted by $E\widehat{\otimes}_{\min} F$, $E\widehat{\otimes}_{\text{proj}} F$, and $E\widehat{\otimes}_h F$.

1.2.1 The Minimal Operator Space Tensor Norm

The *operator space injective tensor norm* (also known as *minimal tensor norm*) arises naturally when considering inclusions of operator spaces into bounded operators on Hilbert spaces and captures the "smallest" possible operator space structure on a tensor product, consistent with the completely bounded setting. The minimal tensor product $E \otimes_{\min} F$ is given by embedding $E \subseteq \mathcal{B}(H_1)$ and $F \subseteq \mathcal{B}(H_2)$ for some Hilbert spaces H_1 and H_2, and then considering the inclusion

$$E \otimes_{\min} F \subseteq \mathcal{B}(H_1 \otimes_2 H_2),$$

where $H_1 \otimes_2 H_2$ denotes the Hilbert space tensor product, that is, the completion of the algebraic tensor product $H_1 \otimes H_2$ with respect to the inner product defined by

$$\langle x_1 \otimes y_1, x_2 \otimes y_2 \rangle = \langle x_1, x_2 \rangle_{H_1} \cdot \langle y_1, y_2 \rangle_{H_2}.$$

The operator space structure on $E \otimes_{\min} F$ is then induced from this inclusion, that is, the matrix norms on $M_n(E \otimes_{\min} F)$ are given by identifying elements with matrices of operators in $\mathcal{B}(H_1 \otimes_2 H_2)$.

Alternatively, the minimal tensor norm of $u \in M_n(E \otimes F)$ can be defined as

$$\min(u) = \sup \left\{ \|(x' \otimes y')_n(u)\|_{M_{npq}} : \|x'\|_{M_p(E')} \leq 1, \|y'\|_{M_q(F')} \leq 1, \ p, q \in \mathbb{N} \right\}, \tag{1.2.1}$$

which resembles the definition of the injective tensor norm ε in the Banach space setting.

It is well known that the minimal tensor product satisfies the following completely isometric inclusions:

$$E \otimes_{\min} F \subset \mathrm{CB}(E', F) \quad \text{and} \quad E' \otimes_{\min} F \subset \mathrm{CB}(E, F).$$

Note that these inclusions turn into equalities whenever E and F are finite-dimensional. This result links the minimal tensor product to the space of completely bounded maps and highlights its role in duality theory.

The norm min is symmetric and associative, i.e. the following canonical identifications are complete isometries:

$$E \otimes_{\min} F \cong F \otimes_{\min} E \quad \text{and} \quad (E \otimes_{\min} F) \otimes_{\min} G \cong E \otimes_{\min} (F \otimes_{\min} G).$$

One of the remarkable features of the minimal tensor product is that it behaves well with respect to complete isometries: if $E_1 \hookrightarrow E_2$ and $F_1 \hookrightarrow F_2$ are complete isometries, then so is the induced map $E_1 \otimes_{\min} F_1 \hookrightarrow E_2 \otimes_{\min} F_2$. This property is commonly referred to by saying that min is completely injective.

Another relevant property of the injective norm min is that it is uniform, meaning that for $S \in \mathrm{CB}(E_1, E_2)$ and $T \in \mathrm{CB}(F_1, F_2)$ we have

$$\|S \otimes T : E_1 \otimes_{\min} F_1 \to E_2 \otimes_{\min} F_2\|_{\mathrm{cb}} \leq \|S\|_{\mathrm{cb}} \|T\|_{\mathrm{cb}}.$$

1.2.2   The Projective Operator Space Tensor Norm

The *operator space projective tensor norm* of $u \in M_n(E \otimes F)$ is defined as

$$\mathrm{proj}(u) = \inf\{\|a\| \cdot \|x\| \cdot \|y\| \cdot \|b\|\}, \tag{1.2.2}$$

where the infimum is taken over all the representations of $u = a(x \otimes y)b$ with $x \in M_p(E)$, $y \in M_q(F)$, $a \in M_{n,p\cdot q}$, $b \in M_{p\cdot q,n}$ for some $p, q \in \mathbb{N}$. Here, the tensor product $x \otimes y$ is viewed as an element of $M_{pq}(E \otimes F)$ under the identification $M_p(E) \otimes M_q(F) \cong M_{pq}(E \otimes F)$.

This formulation reflects the operator space analog of the classical notion of projective tensor norm, but within the matrix-normed setting. The key feature is that the norm takes into account completely bounded factorizations of u through matrices of E and F, preserving the operator space structure at each matrix level. This norm captures a notion of "maximal" operator space structure on the tensor product.

The norm proj is symmetric and associative, i.e. the following canonical identifications are complete isometries:

$$E \otimes_{\mathrm{proj}} F \cong F \otimes_{\mathrm{proj}} E \quad \text{and} \quad (E \otimes_{\mathrm{proj}} F) \otimes_{\mathrm{proj}} G \cong E \otimes_{\mathrm{proj}} (F \otimes_{\mathrm{proj}} G).$$

A significant structural feature of proj is that it is completely projective; that is, if $E_1 \twoheadrightarrow E_2$ and $F_1 \twoheadrightarrow F_2$ are complete quotient maps, then the induced map

$$E_1 \otimes_{\mathrm{proj}} F_1 \twoheadrightarrow E_2 \otimes_{\mathrm{proj}} F_2$$

is also a complete quotient map.

As it happens with the norm min, the tensor norm proj is uniform; that is, for $S \in \mathrm{CB}(E_1, E_2)$ and $T \in \mathrm{CB}(F_1, F_2)$ we have

$$\left\| S \otimes T : E_1 \otimes_{\mathrm{proj}} F_1 \to E_2 \otimes_{\mathrm{proj}} F_2 \right\|_{\mathrm{cb}} \leq \|S\|_{\mathrm{cb}} \|T\|_{\mathrm{cb}}.$$

The projective tensor product also relates to duality in a natural way through the completely isometric identification

$$(E \otimes_{\mathrm{proj}} F)' = \mathrm{CB}(E, F') \cong \mathrm{JCB}(E \times F, \mathbb{C}).$$

Even if the norm proj is not completely injective it preserves the inclusion of a normed operator space into its completion. Indeed, it can be seen through duality arguments that the canonical injection $E \otimes_{\mathrm{proj}} F \hookrightarrow \overline{E} \otimes_{\mathrm{proj}} \overline{F}$ is completely isometric.

1.2.3 The Haagerup Operator Space Tensor Norm

The *Haagerup tensor norm* of $u \in M_n(E \otimes F)$ is defined as

$$h(u) = \inf\{\|x\| \cdot \|y\| : u = x \odot y, x \in M_{n,r}(E), y \in M_{r,n}(F), r \in \mathbb{N}\}, \qquad (1.2.3)$$

where $\odot$ denotes the standard matrix product. Precisely,

$$x \odot y = \left(\sum_{k=1}^{r} x_{ik} \otimes y_{kj} \right)_{i,j}.$$

This tensor norm plays a central role due to its strong structural and analytic properties. It is particularly well suited for interpolation theory, harmonic analysis, and the theory of multiplicately bounded bilinear maps.

The norm h is not symmetric but it is associative so the following canonical identification is a complete isometry

$$(E \otimes_h F) \otimes_h G \cong E \otimes_h (F \otimes_h G).$$

Being that $E \otimes_h F$ could differ from $F \otimes_h E$ it makes sense to define the transpose h^t of h through the identification

$$E \otimes_{h^t} F \cong F \otimes_h E.$$

The Haagerup tensor norm satisfies

$$E \otimes_{\min} F \subseteq E \otimes_h F \subseteq E \otimes_{\text{proj}} F,$$

where all inclusions are completely contractive. Hence, h acts as an intermediate tensor norm between min and proj. As these norms, the Haageruup tensor norm is uniform; i.e. for $S \in \text{CB}(E_1, E_2)$ and $T \in \text{CB}(F_1, F_2)$ we have

$$\|S \otimes T : E_1 \otimes_h F_1 \to E_2 \otimes_h F_2\|_{\text{cb}} \leq \|S\|_{\text{cb}} \|T\|_{\text{cb}} .$$

An astonishing property of the Haagerup tensor norm is that it behaves well with respect to both complete isometries and complete quotients. On the one hand, if $E_1 \hookrightarrow E_2$ and $F_1 \hookrightarrow F_2$ are complete isometries, then so is the induced map $E_1 \otimes_h F_1 \hookrightarrow E_2 \otimes_h F_2$. On the other hand, if $E_1 \twoheadrightarrow E_2$ and $F_1 \twoheadrightarrow F_2$ are complete quotient maps, then the induced map $E_1 \otimes_h F_1 \twoheadrightarrow E_2 \otimes_h F_2$ is also a complete quotient map. We thus say that h is completely injective and completely projective. Note that this phenomenon cannot occur in the category of Banach spaces: there is no tensor norm that is simultaneously injective and projective [23, Prop. 20.20].

The dual of $E \otimes_h F$ is identified with the space of multiplicately bounded bilinear forms on $E \times F$ defined in (1.1.2), i.e.,

$$(E \otimes_h F)' \cong \text{MB}(E \times F, \mathbb{C}).$$

There is some interplay between this tensor norm and min and proj: if R and C denote the canonical row and column operator space structures on ℓ_2 (see definition in Sect. 1.4 below), then for any operator space E,

$$R \otimes_h E \cong R \otimes_{\text{proj}} E, \quad E \otimes_h C \cong E \otimes_{\text{proj}} C, \quad E \otimes_h R \cong E \otimes_{\min} R,$$

$$C \otimes_h E \cong C \otimes_{\min} E$$

completely isometrically.

1.3   Operator Approximation Property

A normed operator space E is said to have the *operator approximation property* (OAP) [34, Sec. 11.2] if there exists a net of finite rank mappings $T_\eta \in \mathrm{CB}(E, E)$ such that the net $id_\mathcal{K} \otimes T_\eta$ converges pointwise to the identity in $K_\infty(E) = \mathcal{K} \widehat{\otimes}_{\min} E$, where $\mathcal{K}$ denotes the compact operators on ℓ_2. Similarly, E is said to have the *C-completely bounded approximation property* (C-CBAP) [34, p. 205] if there exists a net of finite rank mappings $T_\eta \in \mathrm{CB}(E, E)$ such that $\left\| T_\eta \right\|_{\mathrm{cb}} \leq C$ and $\left\| T_\eta x - x \right\| \to 0$ for every $x \in E$. Note that by [34, Thm. 11.3.3], we also have $\left\| (T_\eta)_n x - x \right\| \to 0$ for every $x \in M_n(E)$. In the particular case $C = 1$, we say that E has the *completely metric approximation property* (CMAP).

There is a notable connection between the OAP and tensor products [34, Thm. 11.2.5]: an operator space E has the OAP if and only if for any operator space F, the canonical identity $E \widehat{\otimes}_{\mathrm{proj}} F \to E \widehat{\otimes}_{\min} F$ is one-to-one.

1.4   Vector-valued Versions of the Schatten Spaces

For a Banach operator space E and $1 \leq p \leq \infty$, in [64] Pisier defines an operator space which is an E-valued version of $S_p(H)$ and will be denoted by $S_p[H; E]$: let $S_\infty[H; E] = S_\infty(H) \widehat{\otimes}_{\min} E$ and $S_1[H; E] = S_1(H) \widehat{\otimes}_{\mathrm{proj}} E$, and once again in the case $1 < p < \infty$ we define $S_p[H; E]$ via complex interpolation between $S_\infty[H; E]$ and $S_1[H; E]$. For simplicity we write $S_p[E]$ instead of $S_p[\ell_2; E]$, and $S_p^n[E]$ instead of $S_p[\ell_2^n; E]$. We now briefly recall some important properties of these spaces, and refer the reader to [64] for a detailed study of them.

First of all, we recall a description of the spaces $S_p^n[E]$ and $S_p[E]$ in terms of Haagerup tensor products. In both M_n and $\mathcal{B}(\ell_2)$ we denote by e_{ij} the matrix units, that is, e_{ij} is the linear map whose matrix representation with respect to the canonical basis has a 1 in the (i, j) position and 0's elsewhere. We define the *row spaces*

$$R_n = \mathrm{span}\{e_{1j} \ : \ 1 \leq j \leq n\} \subseteq M_n, \qquad R = \overline{\mathrm{span}}\{e_{1j} \ ; \ j \in \mathbb{N}\} \subseteq \mathcal{B}(\ell_2),$$

and the *column spaces*

$$C_n = \mathrm{span}\{e_{j1} \ : \ 1 \leq j \leq n\} \subseteq M_n, \qquad C = \overline{\mathrm{span}}\{e_{j1} \ ; \ j \in \mathbb{N}\} \subseteq \mathcal{B}(\ell_2),$$

endowed with their natural operator space structures. Since the map $R_n \to C_n$ (resp. $R \to C$) is an isometry, using this isometry as an identification between the two spaces we can perform complex interpolation between them and define, for any $\theta \in [0, 1]$,

$$R_n(\theta) = \big(R_n, C_n\big)_\theta, \qquad R(\theta) = \big(R, C\big)_\theta$$

A crucial point underlying Pisier's construction is that one can identify completely isometrically [65, Cors. 5.10 and 5.11]

$$S_1^n \otimes_{\text{proj}} E = R_n \otimes_h E \otimes_h C_n, \qquad M_n \otimes_{\min} E = C_n \otimes_h E \otimes_h R_n,$$

respectively

$$S_1 \widehat{\otimes}_{\text{proj}} E = R \widehat{\otimes}_h E \widehat{\otimes}_h C, \qquad S_\infty \widehat{\otimes}_{\min} E = C \widehat{\otimes}_h E \widehat{\otimes}_h R,$$

and since the Haagerup tensor product respects complex interpolation that yields complete isometries [64, Thm. 1.1]

$$S_p^n[E] = R_n(1/p') \otimes_h E \otimes_h R_n(1/p), \qquad S_p[E] = R(1/p')\widehat{\otimes}_h E \widehat{\otimes}_h R(1/p)$$

In particular, the identification above interprets elements of $S_p^n[E]$ as $n \times n$ matrices with entries in E in a very particular way: if we denote by $(e_j)_{j=1}^n$ the canonical basis of $R_n(0)$ (i.e. the one coming from the term-by-term identification of the canonical orthonormal basis $(e_{1j})_{j=1}^n$ of R_n and the canonical orthonormal basis $(e_{j1})_{j=1}^n$ of C_n), then the identification above is the map $R_n(1/p') \otimes_h E \otimes_h R_n(1/p) \to S_p^n \otimes E$ given by

$$\sum_{i,j=1}^n e_i \otimes x_{ij} \otimes e_j \mapsto \sum_{i,j=1}^n e_{ij} \otimes x_{ij}.$$

Similarly, by approximation, an element of $S_p[E]$ can be identified with an infinite matrix of elements of E. We note that these identifications as matrices are consistent with the identifications

$$S_p^n \otimes_{\min} E \subseteq \text{CB}(S_{p'}^n, E), \qquad S_p \otimes_{\min} E \subseteq \text{CB}(S_{p'}, E)$$

where an element u in one of these tensor products is identified with the matrix with entries on E which is given by the image of the matrix units under the linear map associated to u.

For a general Hilbert space H, we denote by H_c the corresponding column Hilbert space (that is, with the operator space structure given by the identification with $B(\mathbb{C}, H)$), and by H_r the corresponding row Hilbert space (that is, with the operator space structure given by the identification $H_r = \mathcal{B}(H', \mathbb{C})$. To describe $S_p[H; E]$ in terms of Haagerup tensor products, the identifications for the endpoints $p = 1, \infty$ are given by

$$S_1(H)\widehat{\otimes}_{\text{proj}} E = H_r \widehat{\otimes}_h E \widehat{\otimes}_h (H')_c, \qquad S_\infty(H)\widehat{\otimes}_{\min} E = H_c \widehat{\otimes}_h E \widehat{\otimes}_h (H')_r,$$

and then one proceeds by interpolation as before. We will not write down the details explicitly.

Since we will be making use of operator spaces that are not necessarily Banach, we need to adapt Pisier's definitions to this setting.

Definition 1.4.1 Let E be a normed operator space and $1 \leq p \leq \infty$. We define $S_p^n[E]$ to be the subspace of $S_p^n[\overline{E}]$ consisting of matrices with entries in E. Similarly, we define $S_p[E]$ to be the subspace of $S_p[\overline{E}]$ consisting of matrices with entries in E, and analogously define $S_p[H; E]$ for any Hilbert space H (that this definition does require a choice of orthonormal basis in H, but all choices yield the same space completely isometrically).

Note that clearly $S_p^n[E]$ (resp. $S_p[E]$) is dense in $S_p^n[\overline{E}]$ (resp. $S_p[\overline{E}]$), so we have $\overline{S_p^n[E]} = S_p^n[\overline{E}]$ (resp. $\overline{S_p[E]} = S_p^n[\overline{E}]$).

Pisier's description of $S_p^n[E]$ as a Haagerup tensor product still holds in the case of normed operator spaces, thanks to the injectivity of the Haagerup tensor product. We state this for future reference.

Theorem 1.4.2 *Let E be a normed operator space E. Then, there is a complete isometry*

$$R_n(1/p') \otimes_h E \otimes_h R_n(1/p) = S_p^n[E],$$

given by the identification $\sum_{i,j=1}^n e_i \otimes x_{ij} \otimes e_j \mapsto \sum_{i,j=1}^n e_{ij} \otimes x_{ij}$.

The following is [64, Cor. 1.2 and Lem. 1.7] and describes how completely bounded maps interact with the vector valued S_p spaces. We remark that by density, the argument works for general operator spaces even if they are not Banach.

Theorem 1.4.3 *Let $1 \leq p \leq \infty$. For a linear map $T : E \to F$ between operator spaces, and $k \in \mathbb{N}$,*

$$\|T : E \to F\|_{\mathrm{cb}} = \left\| Id_{S_p} \otimes T : S_p[E] \to S_p[F] \right\|_{\mathrm{cb}}$$

$$= \left\| Id_{S_p^k} \otimes T : S_p^k[E] \to S_p^k[F] \right\|_{\mathrm{cb}} = \sup_{n \in \mathbb{N}} \left\| Id_{S_p^n} \otimes T : S_p^n[E] \to S_p^n[F] \right\|.$$

Moreover, if $T : E \to F$ is a complete isometry (resp. complete quotient) the same is true for $Id_{S_p} \otimes T : S_p[E] \to S_p[F]$ and $Id_{S_p^k} \otimes T : S_p^k[E] \to S_p^k[F]$.

We now recall the duality for the vector-valued Schatten spaces [64, Cor. 1.8].

Theorem 1.4.4 *For $1 < p \leq \infty$, any Hilbert space H and any operator space E, $S_p[H; E]' = S_{p'}[H; E']$.*

The next two results are [64, Thm. 1.9 and Cor. 1.10], once again we remark they remain valid for general operator spaces.

Theorem 1.4.5 (Fubini for the Schatten Spaces) *Let $1 \leq p \leq \infty$. Let H and K be Hilbert spaces, and let E be an operator space. Then we have completely isometrically*

$$S_{\overline{p}}[H; S_p[K; E]] = S_p[H \otimes_2 K; E] = S_p[K; S_p[H; E]].$$

Theorem 1.4.6 (Minkowski Inequality for the Schatten Spaces) *Let* $1 \leq p \leq q \leq \infty$. *Let H and K be Hilbert spaces. Then we have a complete contraction*

$$S_p[H; S_q(K)] \to S_q[K; S_p(H)].$$

For $a, b \in M_n$ we denote by $M(a, b)$ the two-sided multiplication map $M_n \to M_n$ given by $x \mapsto a \cdot x \cdot b$ where $\cdot$ denotes matrix multiplication (and we use the same notation for their infinite-dimensional versions). Their *cb*-norms between Schatten spaces are calculated in [52, Thm. 2.1].

Theorem 1.4.7 *Let* $1 \leq p, q \leq \infty$ *and* $1/r = |1/p - 1/q|$. *Then for every* $a, b \in M_n$

$$\left\| M(a, b) : S_p^n \to S_q^n \right\|_{\mathrm{cb}} = \|a\|_{S_{2r}^n} \|b\|_{S_{2r}^n}.$$

Note that for any operator space E, and $a, b \in M_n$, the map $M(a, b) \otimes Id_E : M_n(E) \to M_n(E)$ is well-defined, and when applied to $x \in M_n(E)$ its image is the "matrix multiplication" of the matrices a, x, and b. Therefore, sometimes we use the notation $a \cdot x \cdot b$ instead of the more cumbersome $(M(a, b) \otimes Id_E)x$ (and similarly for the infinite dimensional two-sided multiplications).

The following two results generalize [64, Thm. 1.5 and Lem. 1.7], see e.g. [17, Lem. 4.2] or [7, Thm. 4.5]. They provide extremely useful ways of calculating norms in vector-valued S_p spaces, and they remain valid for general operator spaces.

Theorem 1.4.8 *Let* $1 \leq p, q, r \leq \infty$ *such that* $1/p = 1/q + 1/r$. *Let* $x \in S_p[E]$ *(resp. $x \in S_p^n[E]$) and let $(x_{ij}) \in M_\infty(E)$ (resp. $(x_{ij}) \in M_n(E)$) be the corresponding matrix with $x_{ij} \in E$. Then $\|x\|_{S_p[E]}$ (resp. $\|x\|_{S_p^n[E]}$) is equal to* $\inf \left\{ \|a\|_{S_{2r}} \|y\|_{S_q[E]} \|b\|_{S_{2r}} \right\}$ *where the infimum runs over all representations of the form $(x_{ij}) = a \cdot y \cdot b$ with $a, b \in S_{2r}$ and $y \in S_q[E]$ (resp. $a, b \in S_{2r}^n$ and $y \in M_n(E)$). Moreover, we can restrict to $a, b \geq 0$.*

Theorem 1.4.9 *Let* $1 \leq p, q, r \leq \infty$ *such that* $1/p = 1/q + 1/r$. *For any* $x \in M_n(E)$,

$$\|x\|_{S_q^n[E]} = \sup\{\|a \cdot x \cdot b\|_{S_p^n[E]} : \|a\|_{S_{2r}^n} \|b\|_{S_{2r}^n} \leq 1\}$$

$$= \sup\{\left\|(M(a, b) \otimes Id_E)x\right\|_{S_p^n[E]} : \|a\|_{S_{2r}^n} \|b\|_{S_{2r}^n} \leq 1\}.$$

While the previous two results appear to only deal with the norm on $S_p^n[E]$, it is easy to adapt them to get versions for the matrix levels $M_m(S_p^n[E])$.

The following "tensor shuffle" style lemma is a generalization of [20, Lem. 3.2]. Note that in particular it implies that the formal identity $S_p^n[E] \to S_p^n \otimes_{\min} E$ (resp. $S_p[E] \to S_p \widehat{\otimes}_{\min} E$) is completely contractive.

Lemma 1.4.10 *For any operator space E, any $1 \leq p, q \leq \infty$ and any $k, m \in \mathbb{N}$, the identity map*

$$S_q^m[S_p^k \otimes_{\min} E] \to S_q^m[S_p^k] \otimes_{\min} E$$

is a complete contraction.

Proof Recall that for $\theta \in [0, 1]$ and $m \in \mathbb{N}$ we denote $R_m(\theta) = (R_m, C_m)_\theta$, where R_m and C_m are the m-dimensional row and column operator spaces, respectively. By Theorem 1.4.2 we have the completely isometric identifications

$$S_q^m[S_p^k \otimes_{\min} E] = R_m(1/q') \otimes_h \left(\left(R_k(1/p') \otimes_h R_k(1/p) \right) \otimes_{\min} E \right) \otimes_h R_m(1/q)$$

and

$$S_q^m[S_p^k] \otimes_{\min} E = \left(R_m(1/q') \otimes_h \left(R_k(1/p') \otimes_h R_k(1/p) \right) \otimes_h R_m(1/q) \right) \otimes_{\min} E.$$

By the "tensor shuffle" [65, Thm. 5.15] and the associativity of the Haagerup tensor product, the shuffle map

$$R_m(1/q') \otimes_h \left(\left(R_k(1/p') \otimes_h R_k(1/p) \right) \otimes_{\min} E \right)$$

$$\to \left(R_m(1/q') \otimes_h R_k(1/p') \otimes_h R_k(1/p) \right) \otimes_{\min} E$$

is a complete contraction. By the uniformity of the Haagerup tensor product, tensoring on the right with $R_m(1/q)$ yields the complete contraction

$$R_m(1/q') \otimes_h \left(\left(R_k(1/p') \otimes_h R_k(1/p) \right) \otimes_{\min} E \right) \otimes_h R_m(1/q)$$

$$\to \left(\left(R_m(1/q') \otimes_h R_k(1/p') \otimes_h R_k(1/p) \right) \otimes_{\min} E \right) \otimes_h R_m(1/q).$$

On the other hand, once again by the "tensor shuffle" and the associativity of the Haagerup tensor product, the shuffle map

$$\left(\left(R_m(1/q') \otimes_h R_k(1/p') \otimes_h R_k(1/p) \right) \otimes_{\min} E \right) \otimes_h R_m(1/q)$$

$$\to \left(R_m(1/q') \otimes_h R_k(1/p') \otimes_h R_k(1/p) \otimes_h R_m(1/q) \right) \otimes_{\min} E$$

is also a complete contraction. Composing the last two maps above yields the desired result because of the identifications at the beginning of the proof. $\qquad\square$

1.5   Relevant Differences with the Banach Space Setting

It is well-known that the theory of Banach spaces cannot be painlessly transferred to the operator space framework, and the theory of tensor norms is no exception. Let us point out three important differences in this context.

1.5.1   Local Reflexivity

A normed operator space E is said to be *locally reflexive* [37] if for each finite-dimensional operator space F, any complete contraction $\varphi : F \to E''$ may be approximated in the point-weak* topology by a net of complete contractions $\varphi_\eta :$ $F \to E$. The class of all locally reflexive normed operator spaces is denoted by OLOC.

Since local reflexivity is relevant to prove many properties related to Banach space tensor norms, it is often an additional hypothesis in the operator space versions of those results. Moreover, in some cases this is known to be necessary: [34, Thm. 14.3.1] is an example of an identity involving tensor norms whose validity turns out to characterize local reflexivity. Nevertheless, there are some particular situations where we can omit this assumption (see, for instance, the results of Chap. 4).

We point out, for future reference, that given a complete operator space E, the completely projective space Z_E introduced in Sect. 1.1.3 is in fact locally reflexive. Indeed, its dual $\ell_\infty(\{M_{n_i} : i \in I\})$ is clearly a C^*-algebra and thus a von Neumann algebra, and preduals of von Neumann algebras are locally reflexive (see e.g., [65, Thm. 18.7]).

1.5.2   Exactness

Every finite-dimensional Banach space embeds $(1 + \varepsilon)$-isomorphically into a space of the form ℓ_∞^n, but the analogous property in the operator space setting does not hold in general: an operator space E is said to be *C-exact* if for any $\varepsilon > 0$ every finite-dimensional subspace of E is $(C + \varepsilon)$-completely isomorphic to a subspace of an M_n space. When an operator space is 1-exact we simply say that it is exact. The comments after the statement of Proposition 10.4.1 (along with Example 10.4.3) give an instance where this causes a significant difference between the classical tensor norm theory and the one for operator spaces. Namely, a simple test to determine whether a tensor norm is the injective associate of a given tensor norm, based on a certain "local" equivalence, does not extend to the setting of operator space tensor norms. At first glance, this might seem unexpected from a naïve Banach space perspective; however, in the non-commutative context, it is well-known that exactness plays a fundamental role in the theory.

Another useful property (for Banach tensor products arguments) which is not transferable to operator spaces due to exactness issues is the one proved in [27, Lem. 2.2.2] about finite dimensional pieces of a quotient:

Theorem 1.5.1 *The following statement is false:*

> *For every complete quotient between Banach operator spaces $q : E \twoheadrightarrow F$, every $F_0 \in \mathrm{OFIN}(F)$, and every $\varepsilon > 0$, there exists $E_0 \in \mathrm{OFIN}(E)$ such that $q(E_0) = F_0$ and for every $y \in M_n(F_0)$ there exists $x \in M_n(E_0)$ such that $q_n(x) = y$ and $\|x\|_{M_n(E_0)} \leq (1 + \varepsilon) \|y\|_{M_n(F_0)}$.*

Proof First let us remark that if $E_0 \in \mathrm{OFIN}(E)$ satisfies the conditions in the statement, then it is clear that F_0 is $(1 + \varepsilon)$-completely isomorphic to $E_0/\ker(q|_{E_0})$ via the canonical map $E_0/\ker(q|_{E_0}) \to F_0$ induced by $q|_{E_0}$.

Let F_0 be a finite-dimensional operator space. By [65, Cor. 2.12.3], there exists a complete quotient $q : S_1 \twoheadrightarrow F_0$. Given $\varepsilon > 0$, suppose that there exists $E_0 \in \mathrm{OFIN}(S_1)$ satisfying the conditions in the statement.

Since $E_0 \in \mathrm{OFIN}(S_1)$, we can find $G_0 \in \mathrm{OFIN}(S_1)$ such that $E_0 \subseteq G_0$ and G_0 is $(1 + \varepsilon)$-completely isomorphic to S_1^N for some $N \in \mathbb{N}$ using perturbation arguments as in [65, Sec. 2.13]. First, take a basis of E_0 and approximate it by finitely supported matrices. This yields a subspace $E_1 \subset S_1$ which is completely isomorphic to E_0 and naturally sits inside a subspace $G_1 \subset S_1$ that is completely isometric to S_1^N for some N. Perturbing G_1 to "put E_1 back onto E_0" yields the desired G_0.

Because $G_0 \supseteq E_0$, it is clear that G_0 also satisfies the conditions in the statement. This gives that F is $(1 + \varepsilon)^2$-completely isomorphic to a quotient of S_1^N, and therefore F' is $(1 + \varepsilon)^2$-completely isomorphic to a subspace of M_N.

But this is, in general, false: from [63, Thm. 7], for any $n \geq 2$ any complete isomorphism from $\mathrm{Max}(\ell_1^n)$ onto a subspace of an M_N space has constant at least $\frac{n}{2\sqrt{n-1}}$. $\qquad\square$

As one would expect, this idea of approximating quotients using finite-dimensional pieces is useful when investigating projectivity of Banach space tensor norms. For example, it is used in [71, Prop 7.5] to show that a tensor norm is projective when its dual tensor norm is injective. Not only does this argument not work in the operator space setting, but the corresponding result is in fact not true: see Remark 12.1.5 below, where we also point out that a related duality result for completely injective and completely projective hulls claimed without proof in [10] does not hold.

1.5.3   Approximation Properties of Completely Injective Spaces

Each Banach space E isometrically embeds into $\ell_\infty(B_{E'})$, which is an injective space in the Banach space setting. The fact that $\ell_\infty(B_{E'})$ has the approximation property plays a relevant role in many tensor product results in the Banach space framework, especially those related with the Approximation Lemma and completely injective tensor norms. In contrast, for an operator space E the standard complete isometry into a completely injective space is the one into $\mathcal{B}(H)$ for some Hilbert space H. The lack of the approximation property in $\mathcal{B}(H)$ is another important dissimilarity between both frameworks.

Chapter 2
Introduction to Operator Space Tensor Norms

This chapter introduces the central focus of this book: operator space tensor norms. It provides a rigorous definition of this concept, drawing inspiration from the classical Banach space setting. It also explores various examples from the literature that fit within this framework and presents new constructions associated to them.

2.1 Definition of Operator Space Tensor Norms

An *operator space cross-norm* α, on the class ONORM of all normed operator spaces, is an assignment of a normed operator space $E \otimes_\alpha F$ to each pair (E, F) of normed operator spaces, in such a way that $E \otimes_\alpha F$ is the algebraic tensor product $E \otimes F$ together with a matricial norm structure on $E \otimes F$, that we write as α_n or $\|\cdot\|_{\alpha_n}$ at each level n (i.e., in $M_n(E \otimes F)$), and such that

$$\alpha_{nm}(x \otimes y) = \|x\|_{M_n(E)} \cdot \|y\|_{M_m(F)} \text{ for every } x \in M_n(E),\, y \in M_m(F). \quad (2.1.1)$$

This implies that the identity map $E \otimes_{\mathrm{proj}} F \to E \otimes_\alpha F$ is completely contractive [13, Thm. 5.5]. If in addition the identity map $E \otimes_\alpha F \to E \otimes_{\min} F$ is also completely contractive, we say that α is *reasonable*. Note that if both mappings $E \otimes_{\mathrm{proj}} F \to E \otimes_\alpha F$ and $E \otimes_\alpha F \to E \otimes_{\min} F$ are completely contractive then we obviously obtain (2.1.1).

Moreover, an operator space cross-norm α is called *uniform* if additionally it satisfies the *complete metric mapping property*: if $S \in \mathrm{CB}(E_1, E_2)$ and $T \in \mathrm{CB}(F_1, F_2)$, then

$$\|S \otimes T : E_1 \otimes_\alpha F_1 \to E_2 \otimes_\alpha F_2\|_{\mathrm{cb}} \leq \|S\|_{\mathrm{cb}} \|T\|_{\mathrm{cb}}.$$

© The Author(s), under exclusive license to Springer Nature Switzerland AG 2025
J. A. Chávez-Domínguez et al., *Operator Space Tensor Norms*, Lecture Notes
in Mathematics 2379, https://doi.org/10.1007/978-3-031-96209-7_2

An operator space cross-norm on ONORM that is both reasonable and uniform will be called an *operator space tensor norm*, or *o.s. tensor norm* for simplicity. In summary:

Definition 2.1.1 An operator space tensor norm α is an assignment to each $E, F \in$ ONORM of a matricial norm in $E \otimes F$ such that the following two conditions are satisfied:

1. α is reasonable: The canonical mappings $E \otimes_{\mathrm{proj}} F \to E \otimes_\alpha F$ and $E \otimes_\alpha F \to E \otimes_{\mathrm{min}} F$ are completely contractive.
2. α is uniform (a short way of saying that α satisfies the complete metric mapping property): If $S \in \mathrm{CB}(E_1, E_2)$ and $T \in \mathrm{CB}(F_1, F_2)$, then

$$\|S \otimes T : E_1 \otimes_\alpha F_1 \to E_2 \otimes_\alpha F_2\|_{\mathrm{cb}} \le \|S\|_{\mathrm{cb}} \|T\|_{\mathrm{cb}} .$$

As usual, we denote by $E \widehat{\otimes}_\alpha F$ the completion of the tensor product $E \otimes_\alpha F$. We remark that in some cases, it will be more natural to define the tensor norm directly in the completed space instead of starting just with the algebraic tensor product.

A degree of caution is required when consulting different works dealing with operator space tensor products, since the term "tensor norm" is not always taken to have the exact same meaning. We are using the same definition as [28, 34], which is slightly different than the one from [10, 13]; we refer to [76, Sec. 6.1] for an explicit comparison. It is clear that analogous definitions for operator space tensor norms can also be made on a subclass of operator spaces, for example the class OFIN of all finite-dimensional operator spaces, or the class of dual operator spaces.

Although we use the notation α_n to refer to an o.s. tensor norm evaluated at level n (that is, on $n \times n$ matrices), we have not adopted this convention before in Sect. 1.2, where we have followed the standard notation in the literature for the classical tensor norms and omitted the subscript.

2.1.1 Basic Properties of Operator Space Tensor Norms

The *transpose* α^t of α is the o.s. tensor norm given by the identification $E \otimes_{\alpha^t} F \cong F \otimes_\alpha E$. An o.s. tensor norm α is *symmetric* if $E \otimes_\alpha F$ and $F \otimes_\alpha E$ are canonically completely isometric (via the transposition map), i.e., $\alpha = \alpha^t$. A norm α is *associative* if $(E \otimes_\alpha F) \otimes_\alpha G$ and $E \otimes_\alpha (F \otimes_\alpha G)$ are canonically completely isometric (via the identity map). As mentioned in Sect. 1.2, min and proj are both symmetric and associative, while the Haagerup o.s. tensor norm h is associative but not symmetric.

An o.s. tensor norm α is called *completely left projective* (resp. *completely left injective*) if for any normed operator space G and any complete projection (resp. complete injection) $T : E \to F$, the map $T \otimes id_G : E \otimes_\alpha G \to F \otimes_\alpha G$ is a complete projection (resp. complete injection) as well. Completely right projective and completely right injective operator space tensor norms are defined analogously, and an o.s. tensor norm will be called completely projective (resp. completely injective) when it is both completely left and right projective (resp. injective).

For operator space tensor norms α and β and a constant c, we write "$\alpha \leq c\beta$ on $E \otimes F$" to indicate that the identity map $E \otimes_\beta F \to E \otimes_\alpha F$ has cb-norm at most c. If no reference to spaces is made, we mean that the inequality holds for any pair of normed operator spaces.

Clearly, when an o.s. tensor norm α is uniform we have that the bilinear map

$$\otimes : \mathrm{CB}(E_1, E_2) \times \mathrm{CB}(F_1, F_2) \to \mathrm{CB}\left(E_1 \otimes_\alpha F_1, E_2 \otimes_\alpha F_2\right) \tag{2.1.2}$$

mapping (S, T) to $S \otimes T$ is jointly contractive. Note that in the case of min, it is even jointly completely contractive. Considering $S \in M_n(\mathrm{CB}(E_1, F_1)) = \mathrm{CB}(E_1, M_n(F_1))$ and $T \in M_m(\mathrm{CB}(E_2, F_2)) = \mathrm{CB}(E_2, M_m(F_2))$, by the uniformity we get

$$\|S \otimes T\|_{\mathrm{CB}(E_1 \otimes_{\min} F_1, M_n(F_1) \otimes_{\min} M_m(F_2))} \leq \|S\|_{\mathrm{CB}(E_1, M_n(F_1))} \|T\|_{\mathrm{CB}(E_2, M_n(F_2))}.$$

Using the identification

$$M_n(F_1) \otimes_{\min} M_m(F_2) = M_n \otimes_{\min} F_1 \otimes_{\min} M_m \otimes_{\min} F_2$$

$$= M_n \otimes_{\min} M_m \otimes_{\min} F_1 \otimes_{\min} F_2 = M_{nm}(F_1 \otimes_{\min} F_2)$$

we equivalently have

$$\|S \otimes T\|_{M_{nm}(\mathrm{CB}(E_1 \otimes_{\min} F_1, E_2 \otimes_{\min} F_2))} \leq \|S\|_{M_n(\mathrm{CB}(E_1, F_1))} \|T\|_{M_m(\mathrm{CB}(E_2, F_2))}.$$

Other examples of o.s. tensor norms for which (2.1.2) is jointly completely contractive are $\alpha = \mathrm{proj}$ and $\alpha = h$ [13].

The following notion will be very useful for defining o.s. tensor norms. If E, F, and G are normed spaces, we define the *tensor contractions*

$$\tau : (E \otimes G') \otimes (G \otimes F) \to E \otimes F$$

$$\sigma : (E \otimes G) \otimes (G \otimes F) \to E \otimes F$$

via

$$\tau(x \otimes g' \otimes g \otimes y) = \langle g', g \rangle x \otimes y, \qquad \sigma(x \otimes g \otimes g' \otimes y) = \langle g', g \rangle x \otimes y$$

and extending linearly. Note that these are defined only as linear maps, and the word "contraction" in this context refers to the fact that the term $G' \otimes G$ gets "contracted" to a scalar (and not to contractivity with respect to a norm). However, the following lemma gives conditions for these tensor contractions to actually be completely contractive.

Lemma 2.1.2 *Let α be an o.s tensor norm for which the map (2.1.2) is jointly completely contractive. Then for any operator spaces E, F and G the tensor contractions*

$$\tau : (E \otimes_{\min} G') \otimes_{\text{proj}} (G \otimes_{\alpha} F) \to E \otimes_{\alpha} F$$

$$\sigma : (E \otimes_{\min} G) \otimes_{\text{proj}} (G' \otimes_{\alpha} F) \to E \otimes_{\alpha} F$$

are completely contractive.

Proof Since

$$\otimes : \text{CB}(G', E) \times \text{CB}(F, F) \to \text{CB}\left(G' \otimes_{\alpha} F, E \otimes_{\alpha} F\right)$$

is jointly completely contractive, we have that the map

$$\text{CB}(G', E) \to \text{CB}\left(G' \otimes_{\alpha} F, E \otimes_{\alpha} F\right)$$

given by $S \mapsto S \otimes Id_F$ is completely contractive, so restricting to $E \otimes_{\min} G \subset \text{CB}(G', E)$ is still completely contractive. This restriction is precisely the map corresponding to σ under the completely isometric identification

$$CB\big((E \otimes_{\min} G) \otimes_{\text{proj}} (G' \otimes_{\alpha} F), E \otimes_{\alpha} F\big) = CB\big(E \otimes_{\min} G, CB(G' \otimes_{\alpha} F, E \otimes_{\alpha} F)\big).$$

The argument for τ is analogous, just using $E \otimes_{\min} G' \subset \text{CB}(G, E)$.   □

2.2 Examples of Operator Space Tensor Norms

In addition to the three fundamental o.s. tensor norms defined in the previous chapter (injective, projective, Haagerup), a number of other examples have appeared in the literature and we list some of them below. The first three are discussed in [35], where it is shown that they satisfy the complete metric mapping property. To conclude that they are o.s. tensor norms, we then just need to check that they are between min and proj.

2.2.1 The Nuclear Tensor Product

$E \widehat{\otimes}_{\text{nuc}} F$ is defined as the quotient of $E \widehat{\otimes}_{\text{proj}} F$ by the kernel of the canonical identity map $E \widehat{\otimes}_{\text{proj}} F \to E \widehat{\otimes}_{\min} F$. From the definition it is clear that $\min \leq \text{nuc} \leq \text{proj}$, so nuc is reasonable. Note that this definition is naturally connected to the classical tensor product characterization of the OAP, see Sect. 1.3.

2.2.2 *The Symmetrized Haagerup Tensor Products*

Given normed operator spaces E and F and $u \in M_n(E \otimes F)$, we define

$$(h \cap h^t)_n(u; E, F) = \max \left\{ h_n(u; E, F), h_n^t(u; E, F) \right\}.$$

It is easy to see that $h \cap h^t$ is a symmetric o.s. tensor norm, and from the definition it is obvious that there is a completely isometric embedding

$$E \otimes_{h \cap h^t} F \hookrightarrow (E \otimes_h F) \oplus_\infty (E \otimes_{h_t} F)$$

given by $u \mapsto (u, u)$. Since h and h^t are completely injective, it is clear that $h \cap h^t$ is completely injective as well. Notice that if α is a symmetric o.s. tensor norm such that $h \leq \alpha$, by transposing we have $h^t \leq \alpha$ and therefore $h \cap h^t \leq \alpha$. Thus, we call $h \cap h^t$ the minimal symmetrized Haagerup tensor norm. This tensor norm has appeared in [29, 42, 69].

On the other hand, for normed operator spaces E and F and $u \in M_n(E \otimes F)$, we define

$$(h + h^t)_n(u; E, F) = \inf \left\{ \|(v, w)\|_{M_n\left((E \otimes_h F) \oplus_1 (E \otimes_{h^t} F)\right)} : u = v + w \right\}.$$

That is, by definition the mapping

$$q : (E \otimes_h F) \oplus_1 (E \otimes_{h^t} E) \to E \otimes_{h + h^t} F$$

given by $q(u, v) = u + v$ is a complete quotient. Once again, it is easy to check that this defines a symmetric o.s. tensor norm which is completely projective because so is h. The o.s. tensor norm $h \mid h^t$ was introduced in [53], where it was denoted by μ and it was shown that it is neither associative nor completely injective. We will call $h + h^t$ the maximal symmetrized Haagerup tensor norm because of the following property. Suppose that β is a symmetric tensor norm with $\beta \leq h$. Then $\beta \leq h^t$, so that the formal identity maps

$$E \otimes_h F \to E \otimes_\beta F, \quad E \otimes_{h^t} F \to E \otimes_\beta F$$

are complete contractions. Therefore, the map $(v, w) \mapsto v + w$ is a complete contraction

$$(E \otimes_h F) \oplus_1 (E \otimes_{h^t} E) \to E \otimes_\beta F$$

and by the standard properties of quotients [65, Prop. 2.4.1], the identity map $E \otimes_{h + h^t} F \to E \otimes_\beta F$ is a complete contraction as well, that is, $\beta \leq h + h^t$.

2.2.3   The Chevet-Saphar Tensor Products

Inspired by the definition of the Banach space case, for $1 \leq p \leq \infty$, the right and left p-Chevet-Saphar o.s. tensor norms, d_p and g_p respectively, were defined originally in [18] and in a slightly different way in [19]. For completeness and for the benefit of the reader, here we want to make clear not just the equivalence between the aforementioned two points of view, but more importantly the ideas behind the construction.

First of all, note that if $Q : X \to Y$ is a linear surjection from a normed space X to a vector space Y, the expression $\|y\|_Q = \inf\{\|x\| : x \in X, \ Qx = y\}$ clearly defines a seminorm on Y, and the map Q becomes a quotient map in the sense that the open unit ball of X gets mapped onto the open unit ball of Y. We now state the operator space version of this construction:

Lemma 2.2.1 *Let $Q : E \to Y$ be a linear surjection from an operator space E to a vector space Y. For each $n \in \mathbb{N}$ and $y \in M_n(Y)$ define*

$$\|y\|_{n,Q} = \inf\{\|x\|_{M_n(E)} : x \in M_n(E), \ Q_n x = y\}.$$

Then the seminorms $\|\cdot\|_{n,Q}$ satisfy Ruan's axioms (see Definition 1.1.1), and the map Q becomes a complete quotient in the sense that for each $n \in \mathbb{N}$, the open unit ball of $M_n(E)$ gets mapped onto the open unit ball of $M_n(Y)$.

The proof is straightforward so we skip it. Just note that, when checking condition **M1**, it is convenient to observe that it suffices to prove the inequality $\leq$, see [34, Prop. 2.3.6].

We start with a finite version of the Chevet-Saphar o.s. tensor norms.

Proposition 2.2.2 *Let $1 \leq p \leq \infty$ and $k \in \mathbb{N}$. For any operator spaces E and F, define $(d_p^k)_n$ as the sequence of seminorms induced by the procedure of Lemma 2.2.1 on $M_n(E \otimes F)$ by the tensor contraction*

$$q^{d_p,k} : (S_{p'}^k \otimes_{\min} E) \otimes_{\mathrm{proj}} S_p^k[F] \to E \otimes F \tag{2.2.1}$$

given by $q^{d_p,k}\big((x_{ij}) \otimes (y_{ij})\big) = \sum_{i,j=1}^k x_{ij} \otimes y_{ij}$ for each $(x_{ij}) \in S_{p'}^k \otimes_{\min} E$ and $(y_{ij}) \in S_p^k[F]$, and extended by linearity. Then d_p^k is an o.s. tensor norm.

Proof Since the identity map $S_p^k[F] \to S_p^k \otimes_{\min} F$ is a complete contraction by Lemma 1.4.10, the complete metric mapping property of proj yields that the identity map

$$(S_{p'}^k \otimes_{\min} E) \otimes_{\mathrm{proj}} S_p^k[F] \to (S_{p'}^k \otimes_{\min} E) \otimes_{\mathrm{proj}} (S_p^k \otimes_{\min} F)$$

is also a complete contraction. Now, the tensor contraction

$$(S_{p'}^k \otimes_{\min} E) \otimes_{\mathrm{proj}} (S_p^k \otimes_{\min} F) \to E \otimes_{\min} F$$

is a complete contraction by Lemma 2.1.2, and thus by taking a composition the mapping

$$(S_{p'}^k \otimes_{\min} E) \otimes_{\text{proj}} S_p^k[F] \to E \otimes_{\min} F$$

is a complete contraction. Therefore for every $u \in M_n(E \otimes F)$ we have $\|u\|_{M_n(E \otimes_{\min} F)} \leq (d_p^k)_n(u)$, which additionally shows that each $(d_p^k)_n$ is in fact a norm and not just a seminorm.

The map $x \mapsto e_{11} \otimes x$ (resp. $y \mapsto e_{11} \otimes y$) is a completely isometric embedding of E (resp. F) into $S_{p'}^k \otimes_{\min} E$ (resp. $S_p^k[F]$). By uniformity of proj their tensor product is then a complete contraction

$$E \otimes_{\text{proj}} F \to (S_{p'}^k \otimes_{\min} E) \otimes_{\text{proj}} S_p^k[F],$$

showing that for every $u \in M_n(E \otimes F)$ we have $(d_p^k)_n(u) \leq \|u\|_{M_n(E \otimes_{\text{proj}} F)}$.

We have then proved $\min \leq d_p^k \leq \text{proj}$, which shows that d_p^k is a reasonable o.s. cross-norm. We now check that it is uniform. For that, let $S \in \text{CB}(E_1, E_2)$ and $T \in \text{CB}(F_1, F_2)$. Since $\left\| Id_{S_p^k} \otimes T : S_p^k[F_1] \to S_p^k[F_2] \right\|_{\text{cb}} = \|T\|_{\text{cb}}$, from the uniformity of min and proj we get that the cb norm of

$$Id_{S_{p'}^k} \otimes S \otimes Id_{S_p^k} \otimes T : (S_{p'}^k \otimes_{\min} E_1) \otimes_{\text{proj}} S_p^k[F_1] \to (S_{p'}^k \otimes_{\min} E_2) \otimes_{\text{proj}} S_p^k[F_2]$$

is at most $\|S\|_{\text{cb}} \|T\|_{\text{cb}}$. Let us now consider the commutative diagram

$$
\begin{array}{ccc}
(S_{p'}^k \otimes_{\min} E_1) \otimes_{\text{proj}} S_p^k[F_1] & \longrightarrow & (S_{p'}^k \otimes_{\min} E_2) \otimes_{\text{proj}} S_p^k[F_2] \\
\downarrow {\scriptstyle q_1} & & \downarrow {\scriptstyle q_2} \\
E_1 \otimes_{d_p^k} F_1 & \xrightarrow{\;\; S \otimes T \;\;} & E_2 \otimes_{d_p^k} F_2
\end{array}
$$

where the top horizontal arrow is $Id_{S_{p'}^k} \otimes S \otimes Id_{S_p^k} \otimes T$, and the vertical arrows are the canonical tensor contractions (which we are denoting slightly differently here for simplicity), which by the definition of d_p^k are complete quotients. Therefore,

$$\|S \otimes T\|_{\text{cb}} = \|(S \otimes T)q_1\|_{\text{cb}} = \left\| q_2 (Id_{S_{p'}^k} \otimes S \otimes Id_{S_p^k} \otimes T) \right\|_{\text{cb}}$$

$$\leq \|q_2\|_{\text{cb}} \left\| Id_{S_{p'}^k} \otimes S \otimes Id_{S_p^k} \otimes T \right\|_{\text{cb}} \leq \|S\|_{\text{cb}} \|T\|_{\text{cb}}.$$

This concludes the proof. $\square$

Note that as a consequence of the previous argument the mapping

$$q^{d_p,k} : (S_{p'}^k \otimes_{\min} E) \otimes_{\text{proj}} S_p^k[F] \to E \otimes_{d_p^k} F \qquad (2.2.2)$$

is a complete quotient.

The Chevet-Saphar o.s. tensor norms will now be obtained as a limit of the finite versions above.

Proposition 2.2.3 *Let* $1 \le p \le \infty$. *For operator spaces* E *and* F, $n \in \mathbb{N}$ *and* $u \in M_n(E \otimes F)$, *define*

$$(d_p)_n(u) = \lim_{k \to \infty} (d_p^k)_n(u) = \inf_{k \in \mathbb{N}} (d_p^k)_n(u).$$

Then d_p *is an o.s. tensor norm.*

Proof The canonical completely isometric inclusion with completely contractively complemented range $S_{p'}^k \to S_{p'}^{k+1}$ (resp. $S_p^k \to S_p^{k+1}$) induces a completely isometric inclusion of $S_{p'}^k \otimes_{\min} E$ (resp. $S_p^k[F]$) as a completely contractively complemented subspace of $S_{p'}^{k+1} \otimes_{\min} E$ (resp. $S_p^{k+1}[F]$). Note that these complementation properties allow us to preserve complete isometries when dealing with the projective o.s. tensor norm. So, this yields a completely isometric inclusion

$$(S_{p'}^k \otimes_{\min} E) \otimes_{\text{proj}} (S_p^k \otimes_{\min} F) \to (S_{p'}^{k+1} \otimes_{\min} E) \otimes_{\text{proj}} (S_p^{k+1} \otimes_{\min} F)$$

which shows that for fixed $n \in \mathbb{N}$ the sequence $(d_p^k)_n$ is decreasing in k and therefore the definition of $(d_p)_n$ makes sense (that is, the limit exists and coincides with the infimum). It is routine to check that d_p inherits all the necessary properties to be an o.s. tensor norm from the corresponding ones for the d_p^k. $\qquad\square$

We are now in a position to show that on the completion $E \widehat{\otimes}_{d_p} F$, the operator space structure has a nice description in terms of a tensor contraction on $(S_{p'} \widehat{\otimes}_{\min} E) \widehat{\otimes}_{\text{proj}} S_p[F]$, which is an analogue of the definition of d_p^k (see Eq. (2.2.1)).

Proposition 2.2.4 *Let* $1 \le p \le \infty$ *and let* E, F *be operator spaces.*

(i) *The bilinear map*

$$B^{d_p} : (S_{p'} \widehat{\otimes}_{\min} E) \times \overline{S_p[F]} \to E \widehat{\otimes}_{d_p} F$$

given by $B^{d_p}(x, y) = \lim_{k \to \infty} \sum_{i,j=1}^k x_{ij} \otimes y_{ij}$ *for any* $x = (x_{ij}) \in S_{p'} \widehat{\otimes}_{\min} E$ *and* $y = (y_{ij}) \in S_p[F]$ *is well defined.*

(ii) *The bilinear map* B^{d_p} *induces a complete contraction*

$$q^{d_p} : (S_{p'} \widehat{\otimes}_{\min} E) \widehat{\otimes}_{\text{proj}} \overline{S_p[F]} \to E \widehat{\otimes}_{d_p} F.$$

(iii) q^{d_p} *is in fact a complete quotient.*

Proof

(*i*) For finitely supported matrices x and y with entries in E and F, respectively, the map is certainly well-defined, and moreover one clearly has that $(d_p)_1\big(B^{d_p}(x, y)\big) \leq \|x\|_{S_{p'}\widehat{\otimes}_{\min}E} \|y\|_{S_p[F]}$. For a general x and y, given $k \in \mathbb{N}$ denote by $x[k]$ and $y[k]$ the truncations of x and y, respectively, to the initial $k \times k$ block. All we need to do is show that $\big(B^{d_p}(x[k], y[k])\big)_{k=1}^{\infty}$ is a Cauchy sequence in $E \otimes_{d_p^o} F$. Note that

$$B^{d_p}\big(x[k], y[k]\big) - B^{d_p}\big(x[l], y[l]\big)$$
$$= B^{d_p}\big(x[k] - x[l], y[k]\big) + B^{d_p}\big(x[l], y[k] - y[l]\big)$$

from where

$$(d_p)_1\big(B^{d_p}\big(x[k], y[k]\big) - B^{d_p}\big(x[l], y[l]\big)\big)$$
$$\leq \|x[k] - x[l]\|_{S_{p'}\widehat{\otimes}_{\min}E} \|y[k]\|_{S_p[F]} + \|x[l]\|_{S_{p'}\widehat{\otimes}_{\min}E} \|y[k] - y[l]\|_{S_p[F]}$$
$$\leq \|x[k] - x[l]\|_{S_{p'}\widehat{\otimes}_{\min}E} \|y\|_{S_p[F]} + \|x\|_{S_{p'}\widehat{\otimes}_{\min}E} \|y[k] - y[l]\|_{S_p[F]},$$

hence $\big(B^{d_p}(x[k], y[k])\big)_{k=1}^{\infty}$ is a Cauchy sequence since so are $(x[k])_{k=1}^{\infty}$ (in $S_{p'}\widehat{\otimes}_{\min}E$) and $(y[k])_{k=1}^{\infty}$ (in $S_p[F]$).

(*ii*) By the universal property of the projective tensor product, to conclude that q^{d_p} is a complete contraction it suffices to check that B^{d_p} is jointly completely bounded. Arguing as in the beginning of the proof of Proposition 2.2.3, for each $k \in \mathbb{N}$ the canonical completely isometric inclusion $S_{p'}^k \to S_{p'}$ (resp. $S_p^k \to S_p$) induces a completely isometric inclusion of $S_{p'}^k \otimes_{\min} E$ (resp. $S_p^k[F]$) as a completely contractively complemented subspace of $S_{p'} \otimes_{\min} E$ (resp. $S_p[F]$). Moreover, the union of the subspaces $S_{p'}^k \otimes_{\min} E$ is dense in $S_{p'}\otimes_{\min} E$, and the union of the subspaces $S_p^k[F]$ is dense in $S_p[F]$. Therefore in order to check that B^{d_p} is jointly completely bounded it suffices to check that each of its restrictions

$$B^{d_p,k} : (S_{p'}^k \otimes_{\min} E) \times S_p^k[F] \to E \otimes_{d_p} F$$

is jointly completely bounded. But the linear map

$$(S_{p'}^k \otimes_{\min} E) \otimes S_p^k[F] \to E \otimes F$$

that corresponds to $B^{d_p,k}$ is precisely the tensor contraction $q^{d_p,k}$ in (2.2.1), meaning that $B^{d_p,k}$ is jointly completely bounded if and only if

$$q^{d_p,k} : (S_{p'}^k \otimes_{\min} E) \otimes_{\text{proj}} S_p^k[F] \to E \otimes_{d_p} F$$

is a complete contraction, but this is clear from writing it as a composition of the complete contractions

$$q^{d_p,k} : (S_{p'}^k \otimes_{\min} E) \otimes_{\text{proj}} S_p^k[F] \to E \otimes_{d_p^k} F \quad \text{and} \quad Id : E \otimes_{d_p^k} F \to E \otimes_{d_p} F.$$

(*iii*) The argument relies on the following observation: suppose X and Y are normed spaces, and $q : X \to Y$ is a linear transformation which maps the open unit ball of X onto the open unit ball of Y (that is, a quotient). In particular q is continuous, so it extends uniquely to a continuous linear transformation $\overline{q}$ from the completion $\overline{X}$ of X to the completion $\overline{Y}$ of Y, and $\overline{q}$ is also a quotient map. Indeed, let $y \in \overline{Y}$ with $\|y\| < 1$. Since Y is dense in $\overline{Y}$, we can write $y = \sum_{j=1}^{\infty} y_j$ with $y_j \in Y$ and $\sum_{j=1}^{\infty} \|y_j\| < 1$. We can then find $x_j \in X$ with $q(x_j) = y_j$ and such that $\sum_{j=1}^{\infty} \|x_j\| < 1$. It follows that $\sum_{j=1}^{\infty} x_j$ converges to some $x \in \overline{X}$ with $\|x\| \le \sum_{j=1}^{\infty} \|x_j\| < 1$, and by continuity $\overline{q}(x) = \sum_{j=1}^{\infty} q(x_j) = \sum_{j=1}^{\infty} y_j = y$.

Now fix $n \in \mathbb{N}$, and let us verify that the n-th amplification of

$$q^{d_p} : (S_{p'}\widehat{\otimes}_{\min}E)\widehat{\otimes}_{\text{proj}} S_p[F] \to E\widehat{\otimes}_{d_p} F$$

is a quotient. To do so, we will take $Y = M_n(E \otimes_{d_p} F)$, whose completion is $\overline{Y} = M_n(E\widehat{\otimes}_{d_p}F)$, and

$$X = \bigcup_{k=1}^{\infty} M_n\big((S_{p'}^k \otimes_{\min} E) \otimes_{\text{proj}} S_p^k[F]\big)$$

where each $(S_{p'}^k \otimes_{\min} E) \otimes_{\text{proj}} S_p^k[F]$ has been completely isometrically identified with a subspace of $(S_{p'} \otimes_{\min} E) \otimes_{\text{proj}} S_p[F]$ in the usual manner. Note that for each $k \in \mathbb{N}$, the restriction of q^{d_p} to the subspace $(S_{p'}^k \otimes_{\min} E) \otimes_{\text{proj}} S_p^k[F]$ is, as a linear map taking values in $E \otimes F$, precisely the map $q^{d_p,k}$ in (2.2.1). Therefore, for each $k \in \mathbb{N}$, the image of the open unit ball of $M_n(X)$ under $(q^{d_p})_n$ contains the image of the open unit ball of $M_n((S_{p'}^k \otimes_{\min} E)\otimes_{\text{proj}} S_p^k[F])$ under $(q^{d_p,k})_n$, which is the open unit ball of $M_n(E \otimes_{d_p^k} F)$ (see Eq. (2.2.2).) Taking the union of the latter over k yields the open unit ball of $M_n(E \otimes_{d_p} F)$, showing that the restriction of $(q^{d_p})_n$ to X is a quotient $X \to Y$. By the observation above, noting that $\overline{X} = M_n\big((S_{p'}\widehat{\otimes}_{\min}E)\widehat{\otimes}_{\text{proj}}\overline{S_p[F]}\big)$, we are done.

$\square$

Remark 2.2.5 It follows from Proposition 2.2.4, taking duals and using Theorem 1.4.4, that for $1 < p \le \infty$ we have a completely isometric embedding

$$(E \otimes_{d_p} F)' \hookrightarrow \text{CB}\left(S_{p'} \otimes_{\min} E, S_{p'}[F']\right),$$

where every $T \in (E \otimes_{d_p} F)'$, understood as an element of $\text{CB}(E, F')$, gets mapped to $Id_{S_{p'}} \otimes T$. These maps are precisely the class of completely p'-summing maps

$\Pi_{p'}(E, F')$ defined by Pisier [64, Chap. 5], see also Chap. 7. Explicitly, there is a canonical identification $(E \otimes_{d_p} F)' = \Pi_{p'}(E, F')$.

For the purposes of doing calculations, one may wish for a more explicit formula for $(d_p)_n$. This can easily be done straight from the definition, although the expressions may not be particularly enlightening. The standard description of the operator space projective tensor product, recalling that $(d_p^k)_n$ is the norm on $M_n(E \otimes F)$ induced by the linear surjection (2.2.1) yields that for $u \in M_n(E \otimes F)$

$$(d_p^k)_n(u) = \inf \left\{ \|a\| \|x\|_{M_r(S_{p'}^k \otimes_{\min} E)} \|y\|_{M_s(S_p^k[F])} \|b\| : u = (q^{d_p,k})_n \big(a(x \otimes y)b\big) \right\}$$

where $a \in M_{n,rs}$, $b \in M_{rs,n}$, $x \in M_r(S_{p'}^k \otimes_{\min} E)$, $y \in M_s(S_p^k[F])$ and $k \in \mathbb{N}$ is fixed. Since $(d_p)_n$ is the infimum of the norms $(d_p^k)_n$ (with respect to k), it then follows that

$$(d_p)_n(u) := \inf \left\{ \|a\| \|x\|_{M_r(S_{p'}^k \otimes_{\min} E)} \|y\|_{M_s(S_p^k[F])} \|b\| : u = (q^{d_p,k})_n \big(a(x \otimes y)b\big) \right\}$$

$$(2.2.3)$$

where a, b, x and y are exactly as before but now k ranges over $\mathbb{N}$.

At the level $n = 1$, it is possible to get a much nicer formula analogous to the one in the classical setting (see [23, Sec. 12.7]).

Theorem 2.2.6 *Let* $1 \leq p \leq \infty$, *let* E *and* F *be operator spaces, and let* $u \in E \otimes F$. *Then*

$$(d_p)_1(u) = \inf \left\{ \big\|(x_{ij})\big\|_{S_{p'}^k \otimes_{\min} E} \big\|(y_{ij})\big\|_{S_p^k[F]} : u = \sum_{i,j=1}^{k} x_{ij} \otimes y_{ij} \right\},$$

where the infimum is taken over all such representations of u *and all* $k \in \mathbb{N}$. *Furthermore, for* $v \in E \widehat{\otimes}_{d_p} F$,

$$(d_p)_1(v) = \inf \left\{ \|x\|_{S_{p'} \widehat{\otimes}_{\min} E} \|y\|_{S_p[F]} : v = B^{d_p}(x, y) \right\}.$$

Proof By the definition of d_p as the infimum of the d_p^k, and Pisier's description of the norm on the projective tensor product [64, Prop. 1.15], we have that

$$(d_p)_1(u) = \inf \left\{ \Big\|\big((x_{ij}^{rs})_{ij}\big)_{rs}\Big\|_{S_{p'}^m[S_{p'}^k \otimes_{\min} E]} \Big\|\big((y_{ij}^{rs})_{ij}\big)_{rs}\Big\|_{S_p^m[S_p^k[F]]} : \right.$$

$$\left. u = \sum_{r,s=1}^{m} \sum_{i,j=1}^{k} x_{ij}^{rs} \otimes y_{ij}^{rs} \right\}$$

By the Fubini theorem for the Schatten spaces, $S_p^m[S_p^k[F]] = S_p^{mk}[F]$. The desired result now follows from the fact that the norm of $\left((x_{ij}^{rs})_{ij}\right)_{rs}$ in $S_{p'}^{mk} \otimes_{\min} E$ is smaller than its norm in $S_{p'}^m[S_{p'}^k \otimes_{\min} E]$, which is a consequence of Lemma 1.4.10 together with the Fubini theorem. The conclusion for elements of the completion is obtained via approximation arguments similar to the ones we used earlier in this section (see also [19, Thm. 2.3] for the argument). $\qquad\square$

We now present a relationship between different Chevet-Saphar tensor norms. In [18, Prop 3.6], a stronger version of the theorem below is stated, but only a sketch of the proof is provided. At this point, we do not know whether the result is indeed true, as the sketch does not appear to offer a conclusive argument. We have also explored alternative strategies to prove it, but so far without success.

Theorem 2.2.7 *For* $1 \le p \le q \le \infty$. *For any operator spaces E and F we have*

$$\left\| E \otimes_{d_p} F \to E \otimes_{d_q} F \right\| \le 1.$$

Proof Let $1/p = 1/q + 1/r$. Let $u \in E \otimes F$ with $(d_p)_1(u) < 1$. By Theorem 2.2.6, there exist $x = (x_{ij}) \in S_{p'}^k \otimes_{\min} E$, $y = (y_{ij}) \in S_p^k[F]$ such that

$$\|x\|_{S_{p'}^k \otimes_{\min} E} < 1, \quad \|y\|_{S_p^k[F]} < 1, \quad u = \sum_{i,j=1}^{k} x_{ij} \otimes y_{ij}.$$

By Theorem 1.4.8, we can write $y = a \cdot \hat{y} \cdot b$ with $\|a\|_{S_{2r}^k} < 1$, $\|b\|_{S_{2r}^k} < 1$, $\|\hat{y}\|_{S_q^k[F]} < 1$. Define $\hat{x} = a^t \cdot x \cdot b^t$. Then

$$u = \sum_{i,j=1}^{k} x_{ij} \otimes y_{ij} = \sum_{i,j=1}^{k} \sum_{r,s=1}^{k} x_{ij} \otimes a_{ir} \hat{y}_{rs} b_{sj} = \sum_{r,s=1}^{k} \left(\sum_{i,j=1}^{k} a_{ir} x_{ij} b_{sj} \right) \otimes \hat{y}_{rs}$$

$$= \sum_{r,s=1}^{k} \hat{x}_{ij} \otimes \hat{y}_{rs},$$

which implies

$$(d_q)_1(u) \le \|\hat{x}\|_{S_{q'}^k \otimes_{\min} E} \|\hat{y}\|_{S_q^k[F]}.$$

Noticing that $1/q' = 1/p' + 1/r$ and using Theorem 1.4.7,

$$\|\hat{x}\|_{S_{q'}^k \otimes_{\min} E} = \|M(a^t, b^t)x\|_{S_{q'}^k \otimes_{\min} E} \le \|a^t\|_{S_{2r}^k} \|b^t\|_{S_{2r}^k} \|x\|_{S_{p'}^k \otimes_{\min} E} < 1$$

which implies $(d_q)_1(u) < 1$.

$\qquad\square$

Just as in the Banach space case, when $p = 1$ or $p = \infty$ there are alternative descriptions for d_p. The formula for d_∞ below is the one that appeared in [33, Sec. 5].

Theorem 2.2.8 *The following characterizations hold:*

(a) $d_1 = \mathrm{proj}.$
(b) *Let E, F be operator spaces. For $u \in M_n(E \otimes F)$,*

$$\|u\|_{M_n(E \otimes_{d_\infty} F)}$$

$$= \inf \left\{ \|t\|_{M_n(E \otimes_{\min} S_1^m)} : t \in M_n(E \otimes S_1^m),\, u = (Id_{M_n} \otimes Id_E \otimes \gamma)(t) \right\} \tag{2.2.4}$$

where the infimum is taken over all $\gamma : S_1^m \to F$ with $\|\gamma\|_{\mathrm{cb}} \leq 1$ and all $m \in \mathbb{N}$. Moreover, if we are given a complete quotient $q : S_1(H) \to F$, then $Id_E \otimes q : E \otimes_{\min} S_1(H) \to E \otimes_{d_\infty} F$ is also a complete quotient.

Proof

(a) We already know that the identity $d_1 \leq \mathrm{proj}$, since d_1 is an o.s. tensor norm. Now, for each $k \in \mathbb{N}$ consider the commutative diagram

$$
\begin{array}{ccc}
(S_\infty^k \otimes_{\min} E) \otimes_{\mathrm{proj}} S_1^k[F] & \xrightarrow{\;I_1\;} & (S_\infty^k \otimes_{\min} E) \otimes_{\mathrm{proj}} (S_1^k \otimes_{\mathrm{proj}} F) \\
\Big\downarrow{\scriptstyle q_1} & & \Big\downarrow{\scriptstyle q_2} \\
E \otimes_{d_1^k} F & \xrightarrow[\;I_2\;]{} & E \otimes_{\mathrm{proj}} F
\end{array}
$$

where the vertical arrows are the tensor contractions. Since q_1 is a complete quotient by definition,

$$\|I_2\|_{\mathrm{cb}} = \|I_2 q_1\|_{\mathrm{cb}} = \|q_2 I_1\|_{\mathrm{cb}} \leq \|q_2\|_{\mathrm{cb}} \|I_1\|_{\mathrm{cb}}$$

Note I_1 is a complete contraction since $S_1^k[F] = S_1^k \otimes_{\mathrm{proj}} F$ by definition, and q_2 is a complete contraction by Lemma 2.1.2, showing that $\mathrm{proj} \leq d_1^k$. Taking the infimum over k, we conclude $\mathrm{proj} \leq d_1$.

(b) This follows immediately from Remark 2.2.5 and [33, Cor. 5.5], since both d_∞ and the expression on the right-hand side of (2.2.4) define operator space structures on $E \otimes F$ that are in duality with $\Pi_1(E, F')$ in exactly the same way. Furthermore, according to [33, Cor. 3.3] we have $\Pi_1(E, F') = \big(E \otimes_{\min} S_1(H)/\ker(Id_E \otimes q)\big)'$ which yields a completely isometric inclusion

$$(E \otimes_{d_\infty} F)' = \Pi_1(E, F') = \big(E \otimes_{\min} S_1(H)/\ker(Id_E \otimes q)\big)' \hookrightarrow (E \otimes_{\min} S_1(H))'.$$

This is precisely the adjoint of $Id_E \otimes q : E \otimes_{\min} S_1(H) \to E \otimes_{d_\infty} F$, which must then be a complete quotient.

$\square$

As a consequence of the previous statement about d_∞, we obtain that for any operator space E, $E \otimes_{\min} S_1(H) = E \otimes_{d_\infty} S_1(H)$.

The o.s. tensor norm g_p is defined as the transpose of d_p, that is, for operator spaces E and F we have that $E \otimes_{g_p} F = F \otimes_{d_p} E$ completely isometrically via the flip map. Clearly, everything we have proved so far for d_p has corresponding versions for g_p.

2.2.4 λ-tensor Products

We now discuss λ-o.s. tensor norms , introduced by Defant and Wiesner in [24, 75]. For each $k \in \mathbb{N}$, let

$$B_k^\lambda : M_k \times M_k \to M_{\tau(k)}$$

(where $\tau(k) \in \mathbb{N}$ is a natural number only depending on k) be a bilinear map; we will denote the sequence $(B_k^\lambda)_k$ by λ. Two basic examples to keep in mind are the tensor product $\otimes$ (i.e. when $B_k^\lambda(x, y) = x \otimes y$ for all k) and the matrix product $\odot$ (i.e. when $B_k^\lambda(x, y) = x \odot y$ for all k). We point out that the notation above is slightly different from that of [24, 75], in order to keep consistency in the notation of the present work. For $u \in M_k(E \otimes F)$, define

$$\lambda_k(u) = \inf\{\|a\| \|v_1\| \|v_2\| \|b\|\} \tag{2.2.5}$$

where the infimum is taken over arbitrary decompositions $u = a \otimes_{B_j^\lambda} (v_1, v_2) b$ with $a \in M_{k,\tau(j)}$, $b \in M_{\tau(j),k}$, $v_1 \in M_j(E)$, $v_2 \in M_j(F)$, where $\otimes_{B_j^\lambda} : M_j(E) \times M_j(F) \to M_{\tau(j)}(E \otimes F)$ is the bilinear map given by $(a_1 \otimes x, a_2 \otimes y) \mapsto B_j^\lambda(a_1, a_2) \otimes x \otimes y$. Observe that the case $\lambda = \otimes$ corresponds to the projective tensor norm, and $\lambda = \odot$ yields the Haagerup tensor norm.

In order to guarantee that the norms defined above give an o.s. tensor norm, we will need for λ to satisfy certain technical conditions. First, some notation. We let $e_{ij} := e_{ij}^{[k,l]} \in M_{k,l}$ denote the matrix which is 1 in the (i, j)-th entry and zero elsewhere, $e_{ij}^{[k]} := e_{ij}^{[k,k]}$, $e_i := e_{ii}$, $e_i^{[k]} := e_{ii}^{[k,k]}$ and $e_{ij}^{[k,l]} = 0$ if $(i, j) \notin \{1, \ldots, k\} \times \{1, \ldots, l\}$. As usual we will denote by I_n the identity $n \times n$-matrix.

Proposition 2.2.9 ([24, Prop. 4.1] and [75, Prop. 6.1]) *For any operator spaces E and F the sequence $\lambda_k(\cdot)$ defined above gives an operator space structure on $E \otimes F$, whenever λ satisfies the following conditions:*

(E1) For all $k \in \mathbb{N}$ there exist $p \in \mathbb{N}$ and matrices $S \in M_{k,\tau(p)}$, $T \in M_{\tau(p),k}$, $a_1, \cdots, a_k \in M_p$ such that for all $j_1, j_2 \in \{1, \cdots, k\}$:

$$SB_p^\lambda(a_{j_1}, a_{j_2})T = \begin{cases} e_j^{[k]} & \text{if } j_1 = j_2 = j, \\ 0 & \text{otherwise.} \end{cases}$$

(E2) For all $r, s \in \mathbb{N}$ there exist matrices $P \in M_{\tau(r)+\tau(s),\tau(r+s)}$, with $\|P\| \le 1$ such that for all $(i_k, j_k) \in \{1, \cdots, r\}^2 \cup \{r+1, \cdots, r+s\}^2$ with $k = 1, 2$:

$$PB_{r+s}^\lambda(e_{i_1 j_1}^{[r+s]}, e_{i_2 j_2}^{[r+s]})P^*$$

$$= \mathrm{diag}\Big(B_r^\lambda(e_{i_1 j_1}^{[r]}, e_{i_2, j_2}^{[r]}), B_s^\lambda(e_{(i_1-r)(j_1-r)}^{[s]}, e_{(i_2-r)(j_2-r)}^{[s]})\Big).$$

(E3) $B_1^\lambda(1, 1) = 1$ and $\sup_{k \in \mathbb{N}} \|B_k^\lambda\| < \infty$.

In this case, we denote by $E \otimes_\lambda F$ the corresponding operator space. Moreover, the complete metric mapping property is satisfied.

Note that, in an abuse of notation, we are using the symbol λ to denote both the sequence of bilinear maps $M_k \times M_k \to M_{\tau(k)}$ and the operator space structure induced on the tensor product.

Just as for the projective and Haagerup tensor products, the dual of a λ-tensor product can be identified with a certain space of bilinear forms. For any bilinear map $\phi : E \times F \to W$ where E, F, W are operator spaces, define the bilinear maps $\phi_{B_k^\lambda} : M_k(E) \times M_k(F) \to M_{\tau(k)}(W)$ given on elementary tensors by

$$\phi_{B_k^\lambda}(a_1 \otimes v_1, a_2 \otimes v_2) = B_k^\lambda(a_1, a_2) \otimes \phi(v_1, v_2), \qquad a_1, a_2 \in M_k, v_1 \in E, v_2 \in F.$$

We also define

$$\|\phi\|_{\mathrm{cb},\lambda} = \sup_{k \in \mathbb{N}} \left\{ \left\| \phi_{B_k^\lambda}(x, y) \right\|_{M_{\tau(k)}(W)} : \|x\|_{M_k(E)} \le 1, \|y\|_{M_k(F)} \le 1 \right\}$$

and

$$\mathrm{CB}_\lambda(E \times F; W) = \left\{ \phi : E \times F \to W \text{ bilinear} : \|\phi\|_{\mathrm{cb},\lambda} < \infty \right\}.$$

This space is an operator space with the identification

$$M_k\big(\mathrm{CB}_\lambda(E \times F; W)\big) = \mathrm{CB}_\lambda\big(E \times F; M_k(W)\big).$$

Note that for the cases $\lambda = \otimes$ (projective tensor product) and $\lambda = \odot$ (Haagerup tensor product) we recover the usual dual spaces $\mathrm{CB}_\otimes = \mathrm{JCB}$ and $\mathrm{CB}_\odot = \mathrm{MB}$.

Theorem 2.2.10 ([24, Thm. 6.2]) *If λ satisfies (E1), (E2), and (E3), then the natural identification yields a canonical complete isometry*

$$\mathrm{CB}(E \otimes_\lambda F, W) = \mathrm{CB}_\lambda(E \times F; W)$$

so in particular

$$(E \otimes_\lambda F)' = \mathrm{CB}_\lambda(E \times F; \mathbb{C}) =: \mathrm{CB}_\lambda(E \times F).$$

Let us denote by $\odot^t : M_k \times M_k \to M_k$ the transposition of the usual matrix product, that is, $A \odot^t B = B \odot A$ for $A, B \in M_k$. It is clear that taking $\lambda = \odot^t$, the associated operator space structure on $E \otimes F$ is precisely $E \otimes_{h^t} F$, the transposition of the Haagerup tensor product. We now introduce a new family of Haagerup-style o.s. tensor products which are λ-tensor products, and which can be understood as a sort of "interpolation" between h and h^t. We refer to them as *interpolated Haagerup o.s. tensor norms* For any $\theta \in [0, 1]$ and $k \in \mathbb{N}$, we let $\odot_k^\theta : M_k \times M_k \to M_k$ be $(1 - \theta) \odot + \theta \odot^t$, that is, $\odot_k^\theta (A, B) = (1 - \theta)AB + \theta BA$ for $A, B \in M_k$. It is clear that each $\odot_k^\theta$ is bilinear, and let us now show that they all induce operator space structures on the tensor product.

Proposition 2.2.11 *For each $\theta \in [0, 1]$, $\odot^\theta$ satisfies (E1), (E2) and (E3).*

Proof Clearly $1 \odot 1 = 1 \odot^t 1 = 1$, and $\|\odot_k\| = \|\odot_k^t\| = 1$ (since $\|A \odot B\| \leq \|A\| \|B\|$ for $A, B \in M_k$). Taking a convex combination yields (E3) for $\odot^\theta$.

As in the proof of [75, Prop. 7.6], (E1) is satisfied for $\odot$ by taking $p = k$, $S = T = I_k$, and $a_{ji} = e_{ji}^{[k]}$. Since all the matrices involved are diagonal all the products commute, and it follows that the same choices yield (E1) for $\odot^t$. Taking convex combinations, the same choices once again yield (E1) for $\odot^\theta$. The exact same type of argument works for (E2): this is proved for $\odot$ in [75, Prop. 7.6] using $P = I_{r+s}$. $\qquad\qquad\square$

Proposition 2.2.12 *For each $\theta \in [0, 1]$, $\odot^\theta$ is a cross-norm.*

Proof Let E, F be operator spaces, $x \in M_m(E)$, $y \in M_n(F)$. It is well-known that $x \otimes y = (x \otimes I_n) \odot (I_m \otimes y)$, see e.g. [34, Eqn. (9.1.10)]. The exact same calculation shows $x \otimes y = (x \otimes I_n) \odot^t (I_m \otimes y)$, and therefore $x \otimes y = \odot_{mn}^\theta (x \otimes I_n, I_m \otimes y)$. Thus, by the definition of $\odot_{mn}^\theta$,

$$\odot_{mn}^\theta (x \otimes y) \leq \|x \otimes I_n\|_{M_{mn}(E)} \|I_m \otimes y\|_{M_{mn}(F)} = \|x\|_{M_m(E)} \cdot \|y\|_{M_n(F)} .$$

$$\square$$

In order to conclude that $\odot^\theta$ is an o.s. tensor norm, the only missing ingredient is to check that $\min \leq \odot^\theta$. We prove a more general result.

Proposition 2.2.13 *Suppose that λ satisfies (E1), (E2), (E3), and $\|B_k^\lambda\|_{\mathrm{jcb}} \leq 1$ for each $k \in \mathbb{N}$. Then the identity mapping $E \otimes_\lambda F \to E \otimes_{\min} F$ is a complete contraction.*

Proof By [13, Thm. 5.1], it suffices to show that for every $U \in M_k(E \otimes F)$, $\phi \in M_m(E')$, $\psi \in M_n(F')$ we have

$$\|\langle\langle \phi \otimes \psi, U\rangle\rangle\|_{M_{mnk}} \leq \|\phi\|_{M_m(E')}\|\psi\|_{M_n(F')}\lambda_k(U).$$

Now, by the identification $(E \otimes_\lambda F)' = \mathrm{CB}_\lambda(E \times F)$ we already know that

$$\|\langle\langle \phi \otimes \psi, U\rangle\rangle\|_{M_{mnk}} \leq \|\phi \otimes \psi\|_{M_{mn}(\mathrm{CB}_\lambda(E \times F))}\lambda_k(U).$$

Recalling that $M_{mn}\big(\mathrm{CB}_\lambda(E \times F)\big) = \mathrm{CB}_\lambda\big(E \times F; M_{mn}\big)$, all we need is to show that

$$\|\varphi\|_{\lambda,\mathrm{cb}} \leq \|\phi\|_{M_m(E')}\|\psi\|_{M_n(F')}$$

where $\varphi = \phi \otimes \psi : E \times F \to M_{mn}$; note that here we are interpreting $\phi \in M_m(E') = \mathrm{CB}(E, M_m)$ and $\psi \in M_n(F') = \mathrm{CB}(F, M_n)$.

To calculate the norm of φ in $\mathrm{CB}_\lambda(E \times F; M_{mn})$, take $u \in M_k(E)$ and $v \in M_k(F)$ with $\|u\|_{M_k(E)}, \|v\|_{M_k(F)} \leq 1$. Represent $u = \sum_i a_i \otimes u_i$, $v = \sum_j b_j \otimes v_j$ where $a_i, b_j \in M_k$, $u_i \in E$, $v_j \in F$. Then, by the definition of $\varphi_{B_k^\lambda}$,

$$\varphi_{B_k^\lambda}(u, v) = \sum_{i,j} B_k^\lambda(a_i, b_j) \otimes \varphi(u_i, v_j) = \sum_{i,j} B_k^\lambda(a_i, b_j) \otimes \phi(u_i) \otimes \psi(v_j)$$

$$= \otimes_{B_k^\lambda}\Big(\sum_i a_i \otimes \phi(u_i), \sum_j b_j \otimes \psi(v_j)\Big) = \otimes_{B_k^\lambda}\big((I_k \otimes \phi)u, (I_k \otimes \phi)v\big).$$

Note that the assumption $\big\|B_k^\lambda\big\|_{\mathrm{jcb}} \leq 1$ precisely means that for any $A \in M_k(M_m)$ and $B \in M_k(M_n)$ we have

$$\Big\|\otimes_{B_k^\lambda}(A, B)\Big\|_{M_{\tau(k)mn}} \leq \|A\|_{M_k(M_m)}\|B\|_{M_k(M_n)},$$

so

$$\Big\|\varphi_{B_k^\lambda}(u, v)\Big\| \leq \|(I_k \otimes \phi)u\|_{M_k(M_m)}\|(I_k \otimes \phi)v\|_{M_k(M_n)}$$

$$\leq \|I_k \otimes \phi\|\,\|u\|\,\|I_k \otimes \phi\|\,\|v\| \leq \|\phi\|\,\|\phi\|$$

which yields the desired conclusion. $\qquad\qquad\square$

Theorem 2.2.14 *For each $\theta \in [0, 1]$, $\odot^\theta$ is an o.s. tensor norm.*

Proof This is an immediate consequence of Propositions 2.2.11, 2.2.12 and 2.2.13, where for the latter we use the fact that $\|\odot\|_{\mathrm{jcb}} \leq 1$ and therefore $\big\|\odot_k^\theta\big\|_{\mathrm{jcb}} \leq 1$ $\quad\square$

Throughout the rest of this paper, whenever we consider any λ-tensor norm we will always be assuming that it satisfies (E1), (E2), (E3) and additionally it is an o.s. tensor norm. To emphasize this point, we will call them λ-*o.s. tensor norms*.

Remark 2.2.15 We have previously pointed out that our notion of uniformity in the definition of o.s. tensor norm is weaker than that of [13]. This has the advantage of allowing the $\odot^\theta$ to be covered by the theory: the only assumptions we are aware of which imply that a λ-tensor norm satisfies the uniformity condition in [13] are conditions (W1) and (W2) in [75, Prop. 12.2], but $\odot^\theta$ only satisfies (W2) in the extreme cases $\theta = 0, 1$.

2.2.5 A General Procedure

In Example 2.2.2 we present two norms obtained by procedures applied to the o.s. tensor norms h and h^t. The same method can be implemented to any pair (α, β) of o.s. tensor norms, thus defining, for given normed operator spaces E and F and $u \in M_n(E \otimes F)$,

$$(\alpha \cap \beta)_n(u; E, F) = \max\left\{\alpha_n(u; E, F), \beta_n(u; E, F)\right\}.$$

It is simple to check that $\alpha \cap \beta$ is an o.s. tensor norm. Note that this definition clearly yields a completely isometric embedding

$$E \otimes_{\alpha \cap \beta} F \hookrightarrow (E \otimes_\alpha F) \oplus_\infty (E \otimes_\beta F)$$

given by $u \mapsto (u, u)$. On the other hand, for normed operator spaces E and F and $u \in M_n(E \otimes F)$, we have

$$(\alpha + \beta)_n(u; E, F) = \inf\left\{\|(v, w)\|_{M_n\left((E \otimes_\alpha F) \oplus_1 (E \otimes_\beta F)\right)} : u = v + w\right\}.$$

In other words, the mapping

$$q : (E \otimes_\alpha F) \oplus_1 (E \otimes_\beta F) \to E \otimes_{\alpha + \beta} F$$

given by $q(u, v) = u + v$ is a complete quotient.

To check that this defines an o.s. tensor norm, it is straightforward to verify Ruan's axioms and the metric mapping property. Moreover, since we have $\min \leq \alpha, \beta \leq \mathrm{proj}$ it immediately follows that $\min + \min \leq \alpha + \beta \leq \mathrm{proj} + \mathrm{proj}$, so it suffices to show that for any o.s. tensor norm α we have $\alpha = \alpha + \alpha$. First, if $\iota_1 : E \otimes_\alpha F \to (E \otimes_\alpha F) \oplus_1 (E \otimes_\alpha F)$ (respectively ι_2) is the inclusion into the first (resp. second) coordinate given by $\iota_1(u) = (u, 0)$ (resp. $\iota_2(u) = (0, u)$), this is clearly a completely isometric inclusion and therefore $q\iota_1$ is a complete contraction, and this composition is simply the formal identity $E \otimes_\alpha F \to E \otimes_{\alpha + \alpha} F$. On the other hand, by the properties of complete quotients and ℓ_1-sums, the cb-norm of the formal identity $I : E \otimes_{\alpha+\alpha} F \to E \otimes_{\alpha} F$ is the maximum of the cb-norms of the maps $I q \iota_1$ and $I q \iota_2$, but both of those are plainly the identity of $E \otimes_{\alpha} F$.

2.3   Other Related Tensor Norm Constructions

In our definition of operator tensor norm (Definition 2.1.1), we are only asking to have a norm defined on the algebraic tensor product. This uniquely determines a norm in the completion, so for many purposes it suffices to just work with the algebraic tensor product and then pass to (norm) limits.

In operator algebra theory, particularly when working with von Neumann algebras, it is often important to consider limits with respect to other topologies. For example, the normal spatial tensor product of von Neumann algebras $A \subseteq \mathcal{B}(H)$ and $B \subseteq \mathcal{B}(K)$ is the weak* closure of $A \otimes B$ inside $\mathcal{B}(H \otimes K)$. In order to provide the reader with a fuller picture of the available tensor products in operator space theory, we now present a couple of examples of closely related constructions which do not exactly fit into our definition.

2.3.1   The Extended Haagerup Tensor Product

In [30] it is defined $E \otimes_{\mathrm{eh}} F$ as the space of maps $u : E' \times F' \to \mathbb{C}$ which are normal (i.e. weak* continuous in each variable) and multiplicatively bounded. If we denote the space of such maps by $\mathrm{MB}^{\sigma}(E' \times F', \mathbb{C})$, the matrix norms on $E \otimes_{\mathrm{eh}} F$ are given by the identification $M_n\big(\mathrm{MB}^{\sigma}(E' \times F', \mathbb{C})\big) = \mathrm{MB}^{\sigma}(E' \times F', M_n)$. It then follows from (1.2.1) that $\min \leq \mathrm{eh}$.

The operator space structure on $E \otimes_{\mathrm{eh}} F$ can be described in the same way as (1.2.3) but replacing $r \in \mathbb{N}$ with an arbitrary set [35, Eqn. 5.10], hence the *extended* in the name. This in particular shows $\mathrm{eh} \leq h$, and thus $\mathrm{eh} \leq \mathrm{proj}$. The extended Haagerup tensor product is associative [35, p. 149], and completely injective but not completely projective [35, Lem. 5.4 and Prop. 5.5]. he extended Haagerup tensor product generalizes the earlier weak* Haagerup tensor product introduced by Blecher and Smith [14], and agrees with it completely isometrically when restricted to dual operator spaces.

2.3.2   The Normal Haagerup Tensor Product

This is only defined for dual operator spaces, by $E' \otimes_{\sigma h} F' = (E \otimes_{\mathrm{eh}} F)'$ [30]. Since the identity $E \otimes_{\mathrm{proj}} F \to E \otimes_{\mathrm{eh}} F$ is a complete contraction, dualizing yields a contraction

$$E' \otimes_{\sigma h} F' = (E \otimes_{\text{eh}} F)' \to (E \otimes_{\text{proj}} F)' = \text{CB}(E, F'),$$

which implies that the identity $E' \otimes_{\sigma h} F' \to E' \otimes_{\min} F'$ is a complete contraction. The fact that $\sigma h \leq \text{proj}$ is a complete contraction, i.e. the identity $E' \otimes_{\text{proj}} F' \to (E \otimes_{\text{eh}} F)'$ is a complete contraction, follows easily from (1.2.2) and the description of $E \otimes_{\text{eh}} F$ as normal multiplicatively bounded mappings (see Sect. 2.3.1).

Recall from above that the extended Haagerup tensor product is not completely projective, that is, there exist complete quotients $\pi_j : E_j \to F_j$, $j = 1, 2$ such that $\pi_1 \otimes \pi_2 : E_1 \otimes_{\text{eh}} E_2 \to F_1 \otimes_{\text{eh}} F_2$ is not a complete quotient. By dualizing, we have complete injections $\pi'_j : F'_j \to E'_j$, $j = 1, 2$ such that $\pi'_1 \otimes \pi'_2 : F'_1 \otimes_{\sigma h} F'_2 \to E'_1 \otimes_{\sigma h} E'_2$ is not a complete injection, so the normal Haagerup tensor product is not completely injective.

Chapter 3
Finite and Cofinite Hulls

In this chapter, we explore two fundamental procedures. The first focuses on understanding the behavior of an operator space tensor norm in finite dimensions, while the second deals with quotients by subspaces of finite codimension. In particular, we introduce the concepts of finitely and cofinitely-generated o.s. tensor norms, which parallel the corresponding notions in the Banach space setting. We establish key properties, provide illustrative examples, and demonstrate that the property of being finitely-generated is preserved under the intersection and sum procedures.

3.1 Definitions and Examples

Given an operator space tensor norm α on OFIN, we can use the following two procedures to extend it to the class of all operator spaces. The first procedure, $\overrightarrow{\alpha}$, is the *finite hull* of α, and the second, $\overleftarrow{\alpha}$, is its *cofinite hull*.

Definition 3.1.1 Suppose α is an o.s. tensor norm on OFIN and let E and F be operator spaces, and let $u \in M_n(E \otimes F)$. We define

$$\overrightarrow{\alpha}_n(u; E, F) := \inf\{\alpha_n(u; E_0, F_0): E_0 \in \mathrm{OFIN}(E), F_0 \in \mathrm{OFIN}(F),$$
$$u \in M_n(E_0 \otimes F_0)\},$$

$$\overleftarrow{\alpha}_n(u; E, F) := \sup\Big\{\alpha_n\big((q_K^E \otimes q_L^F)_n(u); E/K, F/L\big):$$
$$K \in \mathrm{OCOFIN}(E), L \in \mathrm{OCOFIN}(F)\Big\}.$$

© The Author(s), under exclusive license to Springer Nature Switzerland AG 2025
J. A. Chávez-Domínguez et al., *Operator Space Tensor Norms*, Lecture Notes
in Mathematics 2379, https://doi.org/10.1007/978-3-031-96209-7_3

An o.s. tensor norm α on ONORM is called *finitely-generated* if $\alpha = \overrightarrow{\alpha}$, and *cofinitely-generated o.s. tensor norm* if $\alpha = \overleftarrow{\alpha}$.

Clearly, if $\alpha \leq c\beta$ it follows that $\overleftarrow{\alpha} \leq c\overleftarrow{\beta}$ and $\overrightarrow{\alpha} \leq c\overrightarrow{\beta}$. Since min and h are completely injective, $\mathrm{min} = \overrightarrow{\mathrm{min}}$ and $h = \overrightarrow{h}$. Observe also that the description of the projective norm as an infimum gives $\mathrm{proj} = \overrightarrow{\mathrm{proj}}$.

For the same reason, any λ-o.s. tensor norm is finitely-generated.

Proposition 3.1.2 *Let α be an o.s. tensor norm on* OFIN. *Then the* finite hull $\overrightarrow{\alpha}$ *of α and the* cofinite hull $\overleftarrow{\alpha}$ *of α are o.s. tensor norms on* ONORM *with*

$$\mathrm{min} \leq \overleftarrow{\alpha} \leq \overrightarrow{\alpha} \leq \mathrm{proj}, \qquad \overleftarrow{\alpha}\,|_{\mathrm{OFIN}} = \overrightarrow{\alpha}\,|_{\mathrm{OFIN}} = \alpha.$$

If α is defined on ONORM *(and not just on* OFIN*) then*

$$\overleftarrow{\alpha} \leq \alpha \leq \overrightarrow{\alpha}.$$

Proof It is routine to verify that $\overleftarrow{\alpha}$ and $\overrightarrow{\alpha}$ satisfy Ruan's axioms. The complete metric mapping property of α yields $\overleftarrow{\alpha} \leq \alpha \leq \overrightarrow{\alpha}$; the remarks before the statement of the Proposition then show $\mathrm{min} \leq \overleftarrow{\alpha} \leq \overrightarrow{\alpha} \leq \mathrm{proj}$. Since it is obvious that α agrees with $\overleftarrow{\alpha}$ and $\overrightarrow{\alpha}$ on finite-dimensional spaces, the only thing left to prove is that both hulls are in fact o.s. tensor norms. But we have already proved that they are between min and proj, so we just have to check the complete metric mapping property.

To that effect, let $S \in \mathrm{CB}(E_1, E_2)$, $T \in \mathrm{CB}(F_1, F_2)$ and $u \in M_n(E_1 \otimes F_1)$. Then, using the complete metric mapping property of α,

$$\overrightarrow{\alpha}_n\big((S \otimes T)_n u; E_2, F_2\big)$$

$$= \inf\Big\{\alpha_n\big((S \otimes T)_n u; E_0^2, F_0^2\big): E_0^2 \in \mathrm{OFIN}(E_2),$$

$$F_0^2 \in \mathrm{OFIN}(F_2), (S \otimes T)_n u \in M_n(E_0^2 \otimes F_0^2)\Big\}$$

$$\leq \inf\Big\{\alpha_n\big((S \otimes T)_n u; SE_0^1, TF_0^1\big):$$

$$E_0^1 \in \mathrm{OFIN}(E_1), F_0^1 \in \mathrm{OFIN}(F_1), u \in M_n(E_0^1 \otimes F_0^1)\Big\}$$

$$\leq \|S\|_{\mathrm{cb}}\|T\|_{\mathrm{cb}}\inf\Big\{\alpha_n\big(u; E_0^1, F_0^1\big):$$

$$E_0^1 \in \mathrm{OFIN}(E_1), F_0^1 \in \mathrm{OFIN}(F_1), u \in M_n(E_0^1 \otimes F_0^1)\Big\}$$

$$= \|S\|_{\mathrm{cb}}\|T\|_{\mathrm{cb}}\,\overrightarrow{\alpha}_n\big(u; E_1, F_1\big),$$

which shows the complete metric mapping property for $\overrightarrow{\alpha}$.

Now, consider $K_0^2 \in \mathrm{OCOFIN}(E_2)$ and $L_0^2 \in \mathrm{OCOFIN}(F_2)$. Observe that $S^{-1}K_0^2 \in \mathrm{OCOFIN}(E_1)$ and $T^{-1}L_0^2 \in \mathrm{OCOFIN}(F_1)$. Moreover, by the basic properties of quotients there exist maps $S_{K_0^2} : E_1/S^{-1}K_0^2 \to E_2/K_0^2$ and $T_{L_0^2} : F_1/T^{-1}L_0^2 \to F_2/L_0^2$ making the following diagrams commutative

$$
\begin{array}{ccc}
E_1 & \xrightarrow{\ S\ } & E_2 \\
\downarrow{\scriptstyle q^{E_1}_{S^{-1}K_0^2}} & & \downarrow{\scriptstyle q^{E_2}_{K_0^2}} \\
E_1/S^{-1}K_0^2 & \xrightarrow[\ S_{K_0^2}\ ]{} & E_2/K_0^2
\end{array}
\qquad
\begin{array}{ccc}
F_1 & \xrightarrow{\ T\ } & F_2 \\
\downarrow{\scriptstyle q^{F_1}_{T^{-1}L_0^2}} & & \downarrow{\scriptstyle q^{F_2}_{L_0^2}} \\
F_1/T^{-1}L_0^2 & \xrightarrow[\ T_{L_0^2}\ ]{} & F_2/L_0^2
\end{array}
$$

and moreover we have that $\left\| S_{K_0^2} \right\|_{\mathrm{cb}} = \left\| q^{E_2}_{K_0^2} S \right\|_{\mathrm{cb}} \leq \| S \|_{\mathrm{cb}}$ and $\left\| T_{L_0^2} \right\|_{\mathrm{cb}} = \left\| q^{F_2}_{L_0^2} T \right\|_{\mathrm{cb}} \leq \| T \|_{\mathrm{cb}}$. Therefore, once again using the complete metric mapping property of α,

$$
\overleftarrow{\alpha}_n\big((S \otimes T)_n u; E_2, F_2\big)
$$

$$
= \sup\Big\{ \alpha_n\big((q^{E_2}_{K_0^2} \otimes q^{F_2}_{L_0^2})_n (S \otimes T)_n(u); E_2/K_0^2, F_2/L_0^2\big) :
$$

$$
K_0^2 \in \mathrm{OCOFIN}(E_2), L_0^2 \in \mathrm{OCOFIN}(F_2) \Big\}
$$

$$
= \sup\Big\{ \alpha_n\big((S_{K_0^2} \otimes T_{L_0^2})_n (q^{E_1}_{S^{-1}K_0^2} \otimes q^{F_1}_{T^{-1}L_0^2})_n(u); E_2/K_0^2, F_2/L_0^2\big) :
$$

$$
K_0^2 \in \mathrm{OCOFIN}(E_2), L_0^2 \in \mathrm{OCOFIN}(F_2) \Big\}
$$

$$
\leq \| S \|_{\mathrm{cb}} \| T \|_{\mathrm{cb}} \sup\Big\{ \alpha_n\big((q^{E_1}_{K_0^1} \otimes q^{F_1}_{L_0^1})_n(u); E_1/K_0^1, F_1/L_0^1\big) :
$$

$$
K_0^1 \in \mathrm{OCOFIN}(E_1), L_0^1 \in \mathrm{OCOFIN}(F_1) \Big\}
$$

$$
= \| S \|_{\mathrm{cb}} \| T \|_{\mathrm{cb}} \overleftarrow{\alpha}_n\big(u; E_1, F_1\big),
$$

which shows the complete metric mapping property for $\overleftarrow{\alpha}$. $\qquad\square$

As a consequence of the previous proposition, since we already observed that $\min = \overrightarrow{\min}$ it follows that $\min = \overrightarrow{\min} = \overleftarrow{\min}$.

We will see also in Remark 5.2.4 that the Haagerup o.s. tensor norm satisfies $h = \overrightarrow{h} = \overleftarrow{h}$.

Next we show that, just as in the classical theory, the projective norm is not cofinitely-generated.

Proposition 3.1.3

(a) Let W be a Banach space and F a normed operator space. Then $\mathrm{Max}(W) \otimes_{\overleftarrow{\mathrm{proj}}} F$ and $W \otimes_{\overleftarrow{\pi}} F$ are isometrically isomorphic as Banach spaces.

(b) $\overleftarrow{\mathrm{proj}} \neq \mathrm{proj}$.

Proof

(a) Let $u \in W \otimes F$. From the definitions,

$$\overleftarrow{\mathrm{proj}}(u; \mathrm{Max}(W), F) = \sup \big\{ \mathrm{proj}\big((q_K^W \otimes q_L^F)(u); \mathrm{Max}(W)/K, F/L\big) :$$
$$K \in \mathrm{OCOFIN}(\mathrm{Max}(W)), L \in \mathrm{OCOFIN}(F) \big\}$$

and

$$\overleftarrow{\pi}(u; W, F) = \sup \big\{ \pi\big((q_K^W \otimes q_L^F)(u); W/K, F/L\big) :$$
$$K \in \mathrm{COFIN}(W), L \in \mathrm{COFIN}(F) \big\}.$$

By [65, Prop. 3.3] we have that $\mathrm{Max}(W/K) = \mathrm{Max}(W)/K$. Therefore, it follows from [12, Prop. 1.5.12.(1)] that

$$\mathrm{proj}\big((q_K^W \otimes q_L^F)(u); \mathrm{Max}(W)/K, F/L\big) = \pi\big((q_K^W \otimes q_L^F)(u); W/K, F/L\big)$$

and thus $\overleftarrow{\mathrm{proj}}(u; \mathrm{Max}(W), F) = \overleftarrow{\pi}(u; W, F)$.

(b) Since $\overleftarrow{\pi} \neq \pi$ [23, 15.6], there exist Banach spaces W and V such that $W \otimes_\pi V$ and $W \otimes_{\overleftarrow{\pi}} V$ are different; it then follows from part *(a)* and [12, Prop. 1.5.12.(1)] that $\mathrm{Max}(W) \otimes_{\overleftarrow{\mathrm{proj}}} \mathrm{Max}(V)$ and $\mathrm{Max}(W) \otimes_{\mathrm{proj}} \mathrm{Max}(V)$ are different (even just as Banach spaces).

$\square$

Remark 3.1.4 The right-finite hull is presented in Definition 10.3.3. The left-finite hull can be defined analogously.

3.1.1   More Examples

Proposition 3.1.5 *For $1 \leq p \leq \infty$, the o.s. tensor norms d_p and g_p are finitely-generated.*

Proof By symmetry, it suffices to prove it for d_p. This is immediate from the description in (2.2.3), since the complete isometries $E_0 \subseteq E$ and $F_0 \subseteq F$ induce for any $k \in \mathbb{N}$ complete isometries $S_{p'}^k \otimes_{\min} E_0 \to S_{p'}^k \otimes_{\min} E$ (by the injectivity of min) and $S_p^k[F_0] \to S_p^k[F]$ (by Theorem 1.4.3). $\square$

Proposition 3.1.6 *The symmetrized Haagerup o.s. tensor norms $h \cap h^t$ and $h + h^t$ are finitely-generated.*

Proof Since h and h^t are completely injective, it is clear that so is $h \cap h^t$ and therefore the latter is finitely-generated.

Consider operator spaces E and F, $u \in M_n(E \otimes F)$, and $\varepsilon > 0$. Let $v, w \in M_n(E \otimes F)$ such that $u = v + w$ and

$$\|(v, w)\|_{M_n\left((E \otimes_h F) \oplus_1 (E \otimes_{h^t} F)\right)} < (h + h^t)_n(u; E, F) + \varepsilon.$$

Let $E_0 \in \mathrm{OFIN}(E)$ and $F_0 \in \mathrm{OFIN}(F)$ such that $v, w \in M_n(E_0 \otimes F_0)$. Since h and h^t are both completely injective, and so are ℓ_1-sums (this is clear from the description of ℓ_1-sums in Sect. 1.1.2 and the injectivity of $\mathcal{B}(H)$), we have that

$$\|(v, w)\|_{M_n\left((E_0 \otimes_h F_0) \oplus_1 (E_0 \otimes_{h^t} F_0)\right)} = \|(v, w)\|_{M_n\left((E \otimes_h F) \oplus_1 (E \otimes_{h^t} F)\right)}.$$

Since $(h + h^t)_n(u; E_0, F_0) \leq \|(v, w)\|_{M_n\left((E_0 \otimes_h F_0) \oplus_1 (E_0 \otimes_{h^t} F_0)\right)}$, combining the inequalities gives $(h + h^t)_n(u; E_0, F_0) < (h + h^t)_n(u; E, F) + \varepsilon$ which implies that $h + h^t$ is finitely-generated. $\qquad\square$

The previous proposition is a particular case of the fact that intersection and sum procedures of o.s. tensor norms preserve being finitely-generated. The previous proof was significantly simpler than the one for the general statement, as it relied on the injectivity of the norms.

Proposition 3.1.7 *If α and β are finitely-generated o.s. tensor norms then $\alpha \cap \beta$ is finitely-generated as well.*

Proof Consider operator spaces E and F, $u \in M_n(E \otimes F)$, and $\varepsilon > 0$. Adding up finite dimensional subspaces we can find $E_0 \in \mathrm{OFIN}(E)$ and $F_0 \in \mathrm{OFIN}(F)$ such that both inequalities hold

$$\alpha_n(u; E_0, F_0) < \alpha_n(u; E, F)(1 + \varepsilon) \quad \text{and} \quad \beta_n(u; E_0, F_0) < \beta_n(u; E, F)(1 + \varepsilon).$$

Thus, $(\alpha \cap \beta)_n(u; E_0, F_0) < (\alpha \cap \beta)_n(u; E, F)(1 + \varepsilon)$ and hence $\alpha \cap \beta$ is finitely-generated. $\qquad\square$

The argument for the sum of o.s. tensor norms is more involved needing some preparatory results.

To address this, recall that a *matrix set* $\mathbf{A} = (A_n)_n$ *over a vector space E* is a collection of subsets $A_n \subseteq M_n(E)$ defined for each $n \in \mathbb{N}$. A standard example of such a matrix family for an operator space E is its closed matrix unit ball $\left(B_{M_n(E)}\right)_n$.

As introduced in [36], a matrix set $\mathbf{A} = (A_n)$ over E is said to be *absolutely matrix convex* if it satisfies the following properties:

(i) For any $x \in A_n$ and $y \in A_m$, the direct sum $x \oplus y$ lies in A_{m+n}.

(ii) For any $x \in A_n$, $a \in M_{m,n}$, and $b \in M_{n,m}$ with $\|a\|$, $\|b\| \leq 1$, the element axb belongs to A_m.

We now prove a variation of the Hahn-Banach separation theorem for absolutely matrix convex sets . It is based on [36, Sec. 4].

Lemma 3.1.8 *Let V be a vector space, and $\mathbf{A} = (A_n)_{n=1}^{\infty}$ an absolutely matrix convex matrix set over V such that A_1 is absorbent. Suppose that there exist $u \in M_n(V)$ and $r > 0$ such that $(u + rA_n) \cap A_n = \varnothing$. Then there exists a linear map $\varphi : V \to M_n$ such that $\|\varphi_n(u)\|_{M_{n^2}} > 1$, and $\|\varphi_m(v)\|_{M_{nm}} \leq 1$ for every $m \in \mathbb{N}$ and every $v \in A_m$.*

Proof The absorbent absolutely convex set $A_1 \subset V$ determines a locally convex vector topology τ_1 on V, that is, the one induced by the seminorm given by the Minkowski functional ρ_1 of A_1. We will denote by W the dual of (V, τ_1) endowed with the $\sigma(W, V)$-topology, which implies that the dual of W is canonically identified with V. Thus, we are in the situation described at the beginning of [36, Sec. 4]: V and W are locally convex topological vector spaces with a duality pairing such that all the continuous functionals on V are given by elements of W and vice versa.

It is not difficult to see that for any n the absolutely convex set $A_n \subset M_n(V)$ is also absorbent (the details are essentially the same as some of the arguments below) and therefore it determines a locally convex vector topology τ_n on $M_n(V)$ (that is, the one induced by the seminorm given by the Minkowski functional ρ_n of A_n). We claim that τ_n coincides with the topology of entrywise τ_1 convergence: $v^\gamma \overset{\tau_n}{\to} v \Leftrightarrow$ for every $1 \leq i, j \leq n$ we have $v_{ij}^\gamma \overset{\tau_1}{\to} v_{ij}$. To see this, fix a vector $v_0 \in V$ and let $\widetilde{v_0}$ be the matrix in $M_n(V)$ having v_0 in the upper left corner and zeroes elsewhere. By the absolute matrix convexity of $\mathbf{A}$ it follows that $\rho_n(\widetilde{v_0}) \leq \rho_1(v_0)$, since the matrices in $M_n(V)$ which have an element of A_1 in the upper left corner and zeroes elsewhere belong to A_n. The absolute matrix convexity of $\mathbf{A}$ also implies that ρ_n is invariant under permutation matrices, so the inequality above is preserved if we replace $\widetilde{v_0}$ by any other $n \times n$ matrix having v_0 in one entry and zeroes in all the other ones. Therefore for any $v \in M_n(V)$ we have $\rho_n(v) \leq \sum_{i,j=1}^{n} \rho_1(v_{ij})$ (which means that convergence in entrywise τ_1 convergence implies convergence in τ_n). On the other hand, the absolute matrix convexity of $\mathbf{A}$ also implies that $\rho_m(a \cdot v \cdot b) \leq \|a\| \rho_n(v) \|b\|$ for $v \in M_n(V)$ and scalar matrices $a \in M_{m,n}$, $b \in M_{n,m}$. Therefore, for each $v \in M_n(V)$ and each $1 \leq i, j \leq n$ we have $\rho_1(v_{ij}) \leq \rho_n(v)$ (which means that τ_n convergence implies entrywise τ_1 convergence). Thus, we have that the dual of $(M_n(V), \tau_n)$ can be identified with $M_n(W)$ as follows:

- If $\varphi : M_n(V) \to \mathbb{C}$ is a τ_n-continuous functional, for each i, j its restriction to the subspace of matrices which vanish outside of the i, j entry defines a linear map $V \to \mathbb{C}$ which is continuous with respect to τ_1, i.e can be identified with an element $w_{ij} \in W$. We then have for each $v \in M_n(V)$, $\varphi(v) = \sum_{i,j} \langle v_{ij}, w_{ij} \rangle$.
- Given an element $w \in W$, note that the map $M_n(V) \to \mathbb{C}$ given by $v \mapsto \langle v_{ij}, v \rangle$ is τ_n continuous.

Moreover, note that the $\sigma(M_n(V), M_n(W))$-topology on $(M_n(V), \tau_n)$ is precisely the topology of entrywise $\sigma(V, W)$-convergence.

As in [36, Sec. 4], the pairing between V and W induces the scalar pairing between $M_n(V)$ and $M_n(W)$ given by

$$\langle v, w \rangle = \sum_{i,j} \langle v_{ij}, w_{ij} \rangle.$$

We will denote by ω_n the weak topology on $M_n(V)$ induced by this pairing (which is called the weak topology on $M_n(V)$ in [36, Sec. 4]). Note that ω_1 is simply the $\sigma(V, W)$-topology. We will now show that ω_n is the topology of entrywise $\sigma(V, W)$-convergence:

- If $v^\gamma \xrightarrow{\omega_n} v$, for every i, j and every $w \in W$ we get $\langle v_{ij}^\gamma, w \rangle \to \langle v_{ij}, w \rangle$, simply by choosing the matrix in $M_n(W)$ having w in the i, j coordinate and zeros elsewhere.
- If v^γ converges to v in the entrywise $\sigma(V, W)$-topology, for every i, j and every $w_{ij} \in W$ we have $\langle v_{ij}^\gamma, w_{ij} \rangle \to \langle v_{ij}, w_{ij} \rangle$. Adding up we conclude $\langle v^\gamma, w \rangle \to \langle v, w \rangle$ for each $w \in M_n(W)$.

Thus, as claimed, we have proved that $v^\gamma \xrightarrow{\omega_n} v$ if and only if for every $1 \leq i, j \leq n$ we have $v_{ij}^\gamma \xrightarrow{\omega_1} v_{ij}$.

From our assumption, we know that u does not belong to the τ_n-closure of A_n. Since A_n is convex, by the classical Mazur theorem its τ_n-closure coincides with its $\sigma(M_n(V), M_n(W))$-closure, which by the observations above is precisely its closure in the ω_n-topology. Now the desired conclusion follows by a Hahn-Banach argument, specifically the proof of [36, Prop. 4.1]. $\qquad\square$

The following theorem, specifically part (c), reinterprets the condition of being finitely-generated in a categorical way involving morphisms. It is analogous to the fact that $T : E \to F$ is a complete quotient if and only if for every $S : F \to G$ we have $\|ST\|_{\mathrm{cb}} = \|S\|_{\mathrm{cb}}$.

Theorem 3.1.9 *Let $E, F \in \mathrm{ONORM}$ and let α be an o.s. tensor norm. The following are equivalent:*

(a) For every $n \in \mathbb{N}$ and every $u \in M_n(E \otimes F)$ we have $\alpha = \overrightarrow{\alpha}$ on $E \otimes F$.

(b) For every $n \in \mathbb{N}$ we have

$$B^o_{M_n(E \otimes_\alpha F)} = \bigcup_{\substack{E_0 \in \mathrm{OFIN}(E) \\ F_0 \in \mathrm{OFIN}(F)}} B^o_{M_n(E_0 \otimes_\alpha F_0)}$$

(c) For every $G \in \mathrm{ONORM}$ and every (not necessarily continuous) linear map $S : E \otimes_\alpha F \to G$ we have

$$\|S\|_{\mathrm{cb}} = \sup \left\{ \left\| S(i^E_{E_0} \otimes i^F_{F_0}) \cdot F_0 \otimes_\alpha F_0 \to G \right\|_{\mathrm{cb}} \cdot \right.$$

$$\left. E_0 \in \mathrm{OFIN}(E), F_0 \in \mathrm{OFIN}(F) \right\}.$$

Proof The equivalence between (a) and (b) is straightforward, and so is the implication $(b) \Rightarrow (c)$. For each $n \in \mathbb{N}$, let

$$A_n = \bigcup_{\substack{E_0 \in \mathrm{OFIN}(E) \\ F_0 \in \mathrm{OFIN}(F)}} B^o_{M_n(E_0 \otimes_\alpha F_0)} \subset M_n(E \otimes F).$$

Let us assume (c), and suppose that there exist $n \in \mathbb{N}$ and $u \in M_n(E \otimes F)$ such that

$$r := \inf \{\alpha_n(u; E_0, F_0): E_0 \in \mathrm{OFIN}(E), F_0 \in \mathrm{OFIN}(F), u \in M_n(E_0 \otimes F_0)\}$$

$$- \alpha_n(u; E, F) > 0$$

Without loss of generality, we may assume $\alpha_n(u; E, F) = 1$. We claim that $(u + rA_n) \cap A_n = \varnothing$. If there is an element v in that intersection, we have that there exist $E_1, E_2 \in \mathrm{OFIN}(E)$ and $F_1, F_2 \in \mathrm{OFIN}(F)$ such that $v - u \in M_n(E_1 \otimes F_1)$, $\alpha_n(v - u; E_1, F_1) < r$, $v \in E_2 \otimes F_2$, and $\alpha_n(v; E_2, F_2) < 1$. Taking $E_0 = E_1 + E_2$, $F_0 = F_1 + F_2$ we get $v - u \in M_n(E_0 \otimes F_0)$, $v \in M_n(E_0 \otimes F_0)$ (so $u \in M_n(E_0 \otimes F_0)$) and by the metric mapping property of α, $\alpha_n(v - u; E_0, F_0) < r$ and $\alpha_n(v; E_0, F_0) < 1$. It follows that

$$\alpha_n(u; E_0, F_0) \leq \alpha_n(u - v; E_0, F_0) + \alpha_n(v; E_0, F_0) < r + 1,$$

a contradiction.

It is clear that $(A_n)_{n=1}^\infty$ is absolutely matrix convex, and that A_1 is absorbent. Thus, by Lemma 3.1.8, we can find a linear map $\varphi : E \otimes F \to M_n$ such that $\|\varphi_n(u)\|_{M_{n^2}} > 1$ but $\|\varphi_m(v)\|_{M_{nm}} \leq 1$ for all $m \in \mathbb{N}$, $E_0 \in \mathrm{OFIN}(E)$, $F_0 \in \mathrm{OFIN}(F)$, and $v \in B^o_{M_m(E_0 \otimes_\alpha F_0)}$. The former implies $\|\varphi\|_{\mathrm{cb}} > 1$ whereas the latter implies

$$\sup \left\{ \left\| \varphi(i^E_{E_0} \otimes i^F_{F_0}) : E_0 \otimes_\alpha F_0 \to M_n \right\|_{\mathrm{cb}} : E_0 \in \mathrm{OFIN}(E), F_0 \in \mathrm{OFIN}(F) \right\}$$

$$= \sup \left\{ \|\varphi_m(v)\|_{M_{nm}} : E_0 \in \mathrm{OFIN}(E), F_0 \in \mathrm{OFIN}(F), v \in B^o_{M_n(E_0 \otimes_\alpha F_0)} \right\} \leq 1,$$

a contradiction. $\square$

Thanks to the categorical characterization of finitely-generated o.s. tensor norms that we just obtained, we can now prove that this property is preserved under sums.

Corollary 3.1.10 *If α and β are finitely-generated o.s. tensor norms then $\alpha + \beta$ is finitely-generated as well.*

Proof Fix $E, F \in \mathrm{ONORM}$. Let us denote by

$$q : (E \otimes_\alpha F) \oplus_1 (E \otimes_\beta F) \to E \otimes_{\alpha+\beta} F$$

the canonical complete 1-quotient given by $q(u, v) = u + v$. Let $G \in$ ONORM and consider a linear map $S : E \otimes_{\alpha+\beta} F \to G$. The argument below will assume $\|S\|_{cb} < \infty$, but it is easy to adjust it to the case $\|S\|_{cb} = \infty$. By the universal properties of the quotient and the ℓ_1-sum,

$$\|S\|_{cb} = \|Sq\|_{cb} = \max\{\|Sq\iota_\alpha\|_{cb}, \|Sq\iota_\beta\|_{cb}\}, \tag{3.1.1}$$

where

$$\iota_\alpha : E \otimes_\alpha F \to (E \otimes_\alpha F) \oplus_1 (E \otimes_\beta F), \quad \iota_\beta : E \otimes_\alpha F \to (E \otimes_\alpha F) \oplus_1 (E \otimes_\beta F)$$

are the canonical inclusions into the first and second component, respectively. Since α and β are finitely-generated, by Theorem 3.1.9, given $\varepsilon > 0$ we can find $E_1, E_2 \in$ OFIN(E), $F_1, F_2 \in$ OFIN(F) such that

$$\left\| Sq\iota_\alpha(i_{E_1}^E \otimes i_{F_1}^F) : E_1 \otimes_\alpha F_1 \to G \right\|_{cb} \geq (1 - \varepsilon)\|Sq\iota_\alpha\|_{cb}$$

$$\left\| Sq\iota_\beta(i_{E_2}^E \otimes i_{F_2}^F) : E_2 \otimes_\beta F_2 \to G \right\|_{cb} \geq (1 - \varepsilon)\|Sq\iota_\beta\|_{cb}.$$

Taking $E_0 = E_1 + E_2$, $F_0 = F_1 + F_2$ by the metric mapping property of α and β we obtain

$$\left\| Sq\iota_\alpha(i_{E_0}^E \otimes i_{F_0}^F) : E_0 \otimes_\alpha F_0 \to G \right\|_{cb} \geq (1 - \varepsilon)\|Sq\iota_\alpha\|_{cb}$$

$$\left\| Sq\iota_\beta(i_{E_0}^E \otimes i_{F_0}^F) : E_0 \otimes_\beta F_0 \to G \right\|_{cb} \geq (1 - \varepsilon)\|Sq\iota_\beta\|_{cb}.$$

Now observe that $q\iota_\alpha(i_{E_0}^E \otimes i_{F_0}^F) : E_0 \otimes F_0 \to E \otimes F$ is simply $i_{E_0}^E \otimes i_{F_0}^F : E_0 \otimes F_0 \to E \otimes F$, and analogously for $q\iota_\beta(i_{E_0}^E \otimes i_{F_0}^F)$, and thus

$$\left\| S(i_{E_0}^E \otimes i_{F_0}^F) : E_0 \otimes_\alpha F_0 \to G \right\|_{cb} \geq (1 - \varepsilon)\|Sq\iota_\alpha\|_{cb}$$

$$\left\| S(i_{E_0}^E \otimes i_{F_0}^F) : E_0 \otimes_\beta F_0 \to G \right\|_{cb} \geq (1 - \varepsilon)\|Sq\iota_\beta\|_{cb}. \tag{3.1.2}$$

Let us now consider the complete 1-quotient

$$q_0 : (E_0 \otimes_\alpha F_0) \oplus_1 (E_0 \otimes_\beta F_0) \to E_0 \otimes_{\alpha+\beta} F_0$$

and the inclusions

$$\iota_{\alpha,0} : E_0 \otimes_\alpha F_0 \to (E_0 \otimes_\alpha F_0) \oplus_1 (E_0 \otimes_\beta F_0),$$

$$\iota_{\beta,0} : E_0 \otimes_\alpha F_0 \to (E_0 \otimes_\alpha F_0) \oplus_1 (E_0 \otimes_\beta F_0),$$

Once again using the universal properties of quotients and the ℓ_1-sum, and noting that $(i_{E_0}^E \otimes i_{F_0}^F)q_0\iota_{\alpha,0} : E_0 \otimes F_0 \to E \otimes F$ is simply $i_{E_0}^E \otimes i_{F_0}^F : E_0 \otimes F_0 \to E \otimes F$, and analogously for $(i_{E_0}^E \otimes i_{F_0}^F)q_0\iota_{\alpha,0}$,

$$\left\| S(i_{E_0}^E \otimes i_{F_0}^F) : E_0 \otimes_{\alpha+\beta} F_0 \to G \right\|_{\mathrm{cb}} = \left\| S(i_{E_0}^E \otimes i_{F_0}^F)q_0 \right\|_{\mathrm{cb}}$$

$$= \max \left\{ \left\| S(i_{E_0}^E \otimes i_{F_0}^F)q_0\iota_{\alpha,0} \right\|_{\mathrm{cb}} , \left\| S(i_{E_0}^E \otimes i_{F_0}^F)q_0\iota_{\beta,0} \right\|_{\mathrm{cb}} \right\}$$

$$= \max \left\{ \left\| S(i_{E_0}^E \otimes i_{F_0}^F) : E_0 \otimes_{\alpha} F_0 \to G \right\|_{\mathrm{cb}} , \left\| S(i_{E_0}^E \otimes i_{F_0}^F) : E_0 \otimes_{\beta} F_0 \to G \right\|_{\mathrm{cb}} \right\}.$$

Together with (3.1.1) and (3.1.2) we thus conclude

$$\left\| S(i_{E_0}^E \otimes i_{F_0}^F) : E_0 \otimes_{\alpha+\beta} F_0 \to G \right\|_{\mathrm{cb}}$$

$$\geq (1 - \varepsilon) \max\{ \|Sq\iota_{\alpha}\|_{\mathrm{cb}} , \|Sq\iota_{\beta}\|_{\mathrm{cb}} \} = (1 - \varepsilon) \|S\|_{\mathrm{cb}} .$$

$$\square$$

We have previously discussed the behavior of the finitely-generated property under the intersection and sum procedures. Regarding the cofinitely-generated property, Remark 5.3.10 shows that it is preserved under intersection. We do not know whether the same holds for the sum procedure.

Chapter 4
The Five Basic Lemmas

In the theory of tensor products of normed spaces, "The Five Basic Lemmas" (see Section 13 in Defant and Floret's book [23]) are rather simple results which turn out to be "basic for the understanding and use of tensor norms". Namely, they are the *Approximation Lemma*, the *Extension Lemma*, the *Embedding Lemma*, the *Density Lemma* and the *Local Technique Lemma*. We present here the analogous results for the operator space setting and also exhibit some applications as example of their potential. Our presentation follows the lines of [23]. Although the proofs are similar to the Banach space case, the operator space nature of our tensor products introduces some difficulties and we have to deal in most of the cases with additional hypotheses of local reflexivity. However, for the newly introduced family of o.s. tensor norms (called extended λ-o.s. tensor norms, see Definition 4.2.3) the conditions about local reflexivity can be avoided.

4.1 The Approximation Lemma

Lemma 4.1.1 *Let α and β be o.s. tensor norms (on* ONORM*), E and F normed operator spaces, $c \geq 1$ and*

$$\alpha \leq c\beta \quad on \quad E \otimes N$$

for cofinally many $N \in$ OFIN(F). If F has the completely bounded approximation property with constant $C \geq 1$, then

$$\alpha \leq Cc\beta \text{ on } E \otimes F.$$

Proof Fix $n \in \mathbb{N}$ and $z \in M_n(E \otimes F)$, and take $\varepsilon > 0$. Since F has the C-completely bounded approximation property, there is a net $(T_\eta)_\eta$ of finite rank operators in CB(F, F) with cb norm bounded by C such that $\left\| (T_\eta)_n x - x \right\| \to 0$ for every

© The Author(s), under exclusive license to Springer Nature Switzerland AG 2025
J. A. Chávez-Domínguez et al., *Operator Space Tensor Norms*, Lecture Notes
in Mathematics 2379, https://doi.org/10.1007/978-3-031-96209-7_4

$x \in M_n(F)$. Therefore $(id_E \otimes T_\eta)_n(z)$ converges to z in the projective norm, and thus also in the α norm. Hence we have $\alpha_n(z - (id_E \otimes T_\eta)_n(z); E, F) < \varepsilon$ for some η large enough. If we take a subspace N containing $T_\eta(E)$ satisfying the hypothesis of the lemma, by the complete metric mapping property of the o.s. tensor norm β we have

$$\alpha_n(z; E, F) \leq \alpha_n(z - (id_E \otimes T_\eta)_n(z); E, F) + \alpha_n((id_E \otimes T_\eta)_n(z); E, F)$$
$$\leq \varepsilon + \alpha_n((id_E \otimes T_\eta)_n(z); E, N)$$
$$\leq \varepsilon + c\beta_n((id_E \otimes T_\eta)_n(z); E, N)$$
$$\leq \varepsilon + c \left\| T_\eta : E \to N \right\|_{\mathrm{cb}} \beta_n(z; E, F)$$
$$\leq \varepsilon + Cc\beta_n(z; E, F).$$

Since this holds for every $\varepsilon > 0$, we have $\alpha_n(z; E, F) \leq Cc\beta_n(z; E, F)$. $\qquad \square$

By Proposition 3.1.2 and the previous lemma we obtain the following.

Corollary 4.1.2 *Let α be an o.s. tensor norm (on* OFIN*), E and F normed operator spaces with the completely bounded approximation property with constants C_E and C_F, respectively. Then*

$$\overleftarrow{\alpha} \leq \overrightarrow{\alpha} \leq C_E C_F \overleftarrow{\alpha} \quad on \quad E \otimes F.$$

In particular, if E and F both have the completely metric approximation property then $\overleftarrow{\alpha} = \overrightarrow{\alpha}$ on $E \otimes F$.

4.2 The Extension Lemma

Every $\varphi \in (E \otimes_{\mathrm{proj}} F)' = \mathrm{CB}(E, F')$ has a canonical extension $\varphi^\wedge \in (E \otimes_{\mathrm{proj}} F'')' = \mathrm{CB}(E, F'')$ which satisfies for every $x \in E$ and $y'' \in F''$ the relation

$$\langle \varphi^\wedge, x \otimes y'' \rangle = \langle L_\varphi x, y'' \rangle_{F', F''},$$

where $L_\varphi \in \mathrm{CB}(E, F')$ is the linear map associated with φ. The Extension Lemma deals with this situation for more general o.s. tensor norms. Unlike in the Banach space case, local reflexivity does not come for free and we are forced to require it. The item (a) of the following statement previously appeared in [28, Lem. 4.3]. We include the proof here for the reader's benefit. Also, the proof of item (b) is inspired by [28, Prop. 3.1] which presents the same argument just for the norm proj.

Lemma 4.2.1 (Right Extension Lemma) *Let α be a o.s. tensor norm (on* ONORM*), E and F be normed operator spaces and let $\varphi \in (E \otimes_{\mathrm{proj}} F)'$. Then*

$$\varphi \in (E \otimes_\alpha F)' \text{ if and only if } \varphi^\wedge \in (E \otimes_\alpha F'')'$$

and the map $\varphi \mapsto \varphi^\wedge$ is a complete isometry from $(E \otimes_\alpha F)'$ to $(E \otimes_\alpha F'')'$ in the following cases:

(a) α is finitely-generated and F is locally reflexive.
(b) α is a λ-o.s. tensor norm.
(c) $\alpha = h \cap h^t$ or $\alpha = h + h^t$.

Proof The complete metric mapping property implies that $id_E \otimes \kappa_F : E \otimes_\alpha F \hookrightarrow E \otimes_\alpha F''$ is a complete contraction, and hence so is the restriction map $(id_E \otimes \kappa_F)' : (E \otimes_\alpha F'')' \to (E \otimes_\alpha F)'$. This shows that for any $(\varphi_{ij}) \in M_m\big((E \otimes_{\mathrm{proj}} F)'\big)$,

$$\big\|(\varphi_{ij})\big\|_{M_m((E \otimes_\alpha F)')} \le \big\|(\varphi_{ij}^\wedge)\big\|_{M_m((E \otimes_\alpha F'')')} ,$$

that is, the map $\varphi \mapsto \varphi^\wedge$ under consideration is always completely expansive so it will suffice to show that it is also completely contractive.

(a) Now consider $n \in \mathbb{N}$, $z_0 \in M_n(E \otimes F'')$, $M \in \mathrm{OFIN}(E)$ and $N \in \mathrm{OFIN}(F'')$ such that $z_0 \in M_n(M \otimes N)$. Since F is locally reflexive, by [34, Lem. 14.3.3] given $\varepsilon > 0$ there exists $R \in \mathrm{CB}(N, F)$ with $\|R\|_{\mathrm{cb}} \le 1+\varepsilon$ such that for every $y'' \in N$, $x \in M$ and $1 \le i, j \le m$

$$\langle y'', L_{\varphi_{ij}} x \rangle_{F'',F'} = \langle R y'', L_{\varphi_{ij}} x \rangle_{F,F'}.$$

This means that

$$\langle \varphi_{ij}^\wedge, x \otimes y'' \rangle = \langle \varphi_{ij}, (id_E \otimes R)(x \otimes y'') \rangle,$$

thus

$$\langle\langle (\varphi_{ij}^\wedge), z_0 \rangle\rangle = \langle\langle (\varphi_{ij}), (id_E \otimes R)z_0 \rangle\rangle$$

and hence

$$\begin{aligned}
\big\|\langle\langle (\varphi_{ij}^\wedge), z_0 \rangle\rangle\big\|_{M_{mn}} &\le \big\|(\varphi_{ij})\big\|_{M_m((E \otimes_\alpha F)')}\, \alpha_n((id_E \otimes R)z_0; E, F) \\
&\le \big\|(\varphi_{ij})\big\|_{M_m((E \otimes_\alpha F)')}\, \|id_E \otimes R\|_{\mathrm{cb}}\, \alpha_n(z_0; E, N) \\
&\le \big\|(\varphi_{ij})\big\|_{M_m((E \otimes_\alpha F)')}\, \|R\|_{\mathrm{cb}}\, \alpha_n(z_0; E, N) \\
&\le \big\|(\varphi_{ij})\big\|_{M_m((E \otimes_\alpha F)')}\, (1 + \varepsilon)\alpha_n(z_0; E, N)
\end{aligned}$$

which implies the result since α is finitely-generated.

(b) Recall that by Theorem 2.2.10 we can identify $(E \otimes_\lambda F)' = \mathrm{CB}_\lambda(E \times F)$, so all we need to show is that the map $\varphi \mapsto \varphi^\wedge$ gives a complete contraction $\mathrm{CB}_\lambda(E \times F) \to \mathrm{CB}_\lambda(E \times F'')$. This is clear from the definition of $\|\cdot\|_{\mathrm{cb},\lambda}$ and the fact that the unit ball of $M_k(F)$ is weak*-dense in that of $M_k(F'') = M_k(F)''$ by Goldstine's theorem.

(c) For $\alpha = h \cap h^t$, we already know from part (b) that the map $\varphi \mapsto \varphi^\wedge$ gives completely isometric embeddings

$$(E \otimes_h F)' \hookrightarrow (E \otimes_h F'')', \qquad (E \otimes_{h^t} F)' \hookrightarrow (E \otimes_{h^t} F'')'.$$

Taking the ℓ_1-sum we get a completely isometric embedding

$$(E \otimes_h F)' \oplus_1 (E \otimes_{h^t} F)' \hookrightarrow (E \otimes_h F'')' \oplus_1 (E \otimes_{h^t} F'')',$$

that is,

$$\left[(E \otimes_h F) \oplus_\infty (E \otimes_{h^t} F) \right]' \to \left[(E \otimes_h F'') \oplus_\infty (E \otimes_{h^t} F'') \right]'$$

Considering now the diagram

$$
\begin{array}{ccc}
\left[(E \otimes_h F) \oplus_\infty (E \otimes_{h^t} F) \right]' & \hookrightarrow & \left[(E \otimes_h F'') \oplus_\infty (E \otimes_{h^t} F'') \right]' \\
\downarrow & & \downarrow \\
(E \otimes_{h \cap h^t} F)' & \longrightarrow & (E \otimes_{h \cap h^t} F'')'
\end{array}
$$

the top horizontal arrow is a completely isometric embedding, whereas the vertical arrows are complete quotients. It then follows that the bottom arrow is a complete contraction, which yields the result.

Now, for $\alpha = h + h^t$, again recall from part (b) that the map $\varphi \mapsto \varphi^\wedge$ gives completely isometric embeddings

$$(E \otimes_h F)' \to (E \otimes_h F'')', \qquad (E \otimes_{h^t} F)' \to (E \otimes_{h^t} F'')'.$$

Taking the ℓ_∞-sum we get a completely isometric embedding

$$(E \otimes_h F)' \oplus_\infty (E \otimes_{h^t} F)' \to (E \otimes_h F'')' \oplus_\infty (E \otimes_{h^t} F'')',$$

that is,

$$\left[(E \otimes_h F) \oplus_1 (E \otimes_{h^t} F) \right]' \to \left[(E \otimes_h F'') \oplus_1 (E \otimes_{h^t} F'') \right]'.$$

Considering now the diagram

$$\left[(E \otimes_h F) \oplus_1 (E \otimes_{h^t} F)\right]' \lhook\joinrel\longrightarrow \left[(E \otimes_h F'') \oplus_1 (E \otimes_{h^t} F'')\right]'$$

$$\uparrow \qquad\qquad\qquad\qquad\qquad\qquad \uparrow$$

$$(E \otimes_{h+h^t} F)' \longrightarrow (E \otimes_{h+h^t} F'')'$$

since the vertical arrows and the top one are completely isometric embeddings, we conclude that the bottom arrow is also a completely isometric embedding.

$\square$

Remark 4.2.2 Note that the proof of part (c) more generally shows that Lemma 4.2.1 holds for the intersection and the sum of two o.s. tensor norms for which the lemma already holds. In particular, it applies to the sum and the intersection of any two λ-o.s. tensor norms. This suggests the following concept.

Definition 4.2.3 We say that an o.s. tensor norm is an *extended λ-o.s. tensor norm* (and write $\mathcal{E}(\lambda)$-o.s. tensor norm for simplicity), if it belongs to the smallest collection of o.s. tensor norms which contains the λ-o.s. tensor norms, and is closed under taking sum and intersection. We emphasize that obviously λ-o.s. tensor norms, $h \cap h^t$ and $h + h^t$ are clear examples of this type of norms. Note that, due to Proposition 3.1.7 and Corollary 3.1.10, each extended λ-o.s. tensor norm is finitely-generated.

In a similar way a left extension for each $\varphi \in (E \otimes_{\text{proj}} F)'$ is defined: let $^\wedge\varphi \in (E'' \otimes_\alpha F)'$ given by, for $x'' \in E''$ and $y \in F$,

$$\langle {}^\wedge\varphi, x'' \otimes y \rangle = \langle x'', R_\varphi y \rangle_{E'',E'},$$

where $R_\varphi \in \text{CB}(F, E')$ is the mapping $R_\varphi(y)(x) = \langle \varphi, x \otimes y \rangle$.

Lemma 4.2.4 (Left Extension Lemma) *Let α be a o.s. tensor norm (on* ONORM*), E and F be normed operator spaces and let $\varphi \in (E \otimes_{\text{proj}} F)'$. Then*

$$\varphi \in (E \otimes_\alpha F)' \text{ if and only if } {}^\wedge\varphi \in (E'' \otimes_\alpha F)'.$$

and the map $\varphi \mapsto {}^\wedge\varphi$ is a complete isometry from $(E \otimes_\alpha F)'$ to $(E'' \otimes_\alpha F)'$ in the following cases:

(a) α is finitely-generated and E is locally reflexive.
(b) α is an $\mathcal{E}(\lambda)$-o.s. tensor norm.

Now, both procedures can be applied to a given $\varphi \in (E \otimes_\alpha F)'$ to obtain two (possibly different) extensions to $(E'' \otimes_\alpha F'')'$.

Corollary 4.2.5 *Let α be a o.s. tensor norm (on* ONORM*), E and F be normed operator spaces and let $\varphi \in (E \otimes_{\text{proj}} F)'$. Then the following maps from $(E \otimes_\alpha F)'$ to $(E'' \otimes_\alpha F'')'$ are complete isometries:*

$$\varphi \mapsto {}^\wedge(\varphi^\wedge) \quad and \quad \varphi \mapsto ({}^\wedge\varphi)^\wedge$$

in the following cases:

(a) α is finitely-generated and E and F are locally reflexive.
(b) α is an $\mathcal{E}(\lambda)$-o.s. tensor norm.

As we have already mentioned, the fact that we do not need local reflexivity in the Extension Lemmas if the o.s. tensor norm involved is proj was proved in [28, Prop. 3.1]. In the same article this fact is also shown for the Haagerup o.s. tensor norm h using that it satisfies the following property: for any operator spaces E and F and $\varphi \in (E \otimes_h F)'$, there is a unique extension $\widetilde{\varphi} \in (E'' \otimes_h F'')'$ which is separately w^*-continuous, see [12, 1.6.7].

A word can be said about the equality/non-equality of the two extensions. Repeating the arguments of [23, Cor. 1.9] we obtain the analogous result in our setting. That is, given $\varphi \in (E \otimes_{\mathrm{proj}} F)'$, we have $^\wedge(\varphi^\wedge) = (^\wedge\varphi)^\wedge$ if and only if $L_\varphi : E \to F'$ is weakly compact. This is also equivalent to the existence of an extension of φ to $(E'' \otimes_{\mathrm{proj}} F'')'$ that is separately w^*-continuous. Note that, by the comment above, if $\varphi \in (E \otimes_h F)'$ then L_φ is weakly compact.

Note that we have the separating duality pair $\big\langle E \otimes F'', (E \otimes_\alpha F)' \big\rangle$ given by

$$\langle z, \varphi \rangle = \varphi^\wedge(z).$$

Using the bipolar theorem it is easy to see that, in the Banach space realm, the unit ball $B_{E \otimes_\alpha F}$ is $\sigma(E \otimes F'', (E \otimes_\alpha F)')$-dense in the unit ball $B_{E \otimes_\alpha F''}$, whenever α is a finitely-generated tensor norm. In other words, given $z \in B_{E \otimes_\alpha F''}$ there is a net $(u_\eta)_\eta \subset B_{E \otimes_\alpha F}$ such that for every $\varphi \in (E \otimes_\alpha F)'$,

$$\varphi(u_\eta) \to \varphi^\wedge(z).$$

In our setting, we can consider for each $n \in \mathbb{N}$, the following duality pair:

$$\big\langle M_n(E \otimes F''), S_1^n[(E \otimes_\alpha F)'] \big\rangle$$

defined as

$$\langle (z_{ij}), (\varphi_{ij}) \rangle = \sum_{i,j} \varphi_{ij}^\wedge(z_{ij}).$$

It is easy to see that this dual pairing system is separating. Moreover, we have the following density result.

Corollary 4.2.6 *Let α be a o.s. tensor norm (on* ONORM*), E and F be normed operator spaces.*

For every $n \in \mathbb{N}$, the unit ball $B_{M_n(E \otimes_\alpha F)}$ is $\sigma(M_n(E \otimes F''), S_1^n[(E \otimes_\alpha F)'])$-dense in the unit ball $B_{M_n(E \otimes_\alpha F'')}$ in the following cases:

(a) α is finitely-generated and F is locally reflexive.
(b) α is an $\mathcal{E}(\lambda)$-o.s. tensor norm.

Proof Since $(M_n(E \otimes_\alpha F))'$ is $S_1^n[(E \otimes_\alpha F)']$ isometrically, it is clear that the polar $\left(B_{M_n(E \otimes_\alpha F)}\right)^\circ$ (in this pairing system) is $B_{S_1^n[(E \otimes_\alpha F)']}$. By the bipolar theorem it is enough to prove that $B_{M_n(E \otimes_\alpha F'')} \subset \left(B_{S_1^n[(E \otimes_\alpha F)']}\right)^\circ$. Indeed, given $(z_{ij}) \in B_{M_n(E \otimes_\alpha F'')}$ and $(\varphi_{ij}) \in B_{S_1^n[(E \otimes_\alpha F)']}$; by the Extension Lemma 4.2.1 along with [34, Thm. 4.1.8] we know that $(\varphi_{ij}^\wedge) \in B_{S_1^n[(E \otimes_\alpha F'')']}$. Using again $\left(M_n(E \otimes_\alpha F'')\right)' = S_1^n[(E \otimes_\alpha F'')']$ isometrically, we derive that

$$\left| \langle (z_{ij}), (\varphi_{ij}) \rangle \right| = \left| \sum_{i,j} \varphi_{ij}^\wedge(z_{ij}) \right| \leq 1.$$

This concludes the proof. $\qquad\qquad\qquad\qquad\qquad\qquad\qquad\qquad\qquad\qquad\square$

4.3   The Embedding Lemma

Operator space tensor norms generally do not respect subspaces, but the embedding into the bidual is preserved under certain conditions.

Lemma 4.3.1 *Let α be an o.s. tensor norm (on* ONORM*). The mapping*

$$id_E \otimes \kappa_F : E \otimes_\alpha F \to E \otimes_\alpha F''$$

is a complete isometry in the following cases:

(a) α is finitely-generated and F is locally reflexive.
(b) α is an $\mathcal{E}(\lambda)$-o.s. tensor norm.
(c) α is cofinitely-generated.

Proof Let $n \in \mathbb{N}$ and $z \in M_n(E \otimes_\alpha F)$. By the complete metric mapping property

$$id_E \otimes \kappa_F : E \otimes_\alpha F \to E \otimes_\alpha F''$$

is always a complete contraction, so

$$\alpha_n(z; E, F'') \leq \alpha_n(z; E, F).$$

(a) and *(b)* By the Extension Lemma,

$$\begin{aligned}
\alpha_n&(z; E, F) \\
&= \sup\left\{ \| \langle\langle z, \Phi \rangle\rangle \|_{M_{nm}} \ : \ \Phi \in M_m((E \otimes_\alpha F)'), \|\Phi\|_{M_m((E \otimes_\alpha F)')} \leq 1 \right\} \\
&= \sup\left\{ \| \langle\langle z, \Phi^\wedge \rangle\rangle \|_{M_{nm}} \ : \ \Phi \in M_m((E \otimes_\alpha F)'), \|\Phi\|_{M_m((E \otimes_\alpha F)')} \leq 1 \right\} \\
&\leq \sup\left\{ \| \langle\langle z, \Psi \rangle\rangle \|_{M_{nm}} \ : \ \Psi \in M_m((E \otimes_\alpha F'')'), \|\Psi\|_{M_m((E \otimes_\alpha F'')')} \leq 1 \right\} \\
&= \alpha_n(z; E, F''),
\end{aligned}$$

which gives the reverse inequality.

(c) Let $L \in \mathrm{OCOFIN}(F)$, then L^{00} (formed in F'') is in $\mathrm{OCOFIN}(F'')$ and the map

$$\kappa_{F/L} : F/L \to (F/L)'' = F''/L^{00}$$

is completely isometric and surjective; moreover, $q_{L^{00}}^{F''} \circ \kappa_F = \kappa_{F/L} \circ q_L^F$. Therefore,

$$\alpha_n\big((id_E \otimes q_L^F)_n(z); E, F/L\big) = \alpha_n\big((id_E \otimes (\kappa_{F/L} \circ q_L^F))_n(z); E, (F/L)''\big)$$

$$= \alpha_n\big((id_E \otimes q_{L^{00}}^{F''}) \circ (id_E \otimes \kappa_F))_n(z); E, F''/L^{00}\big)$$

$$\leq \overleftarrow{\alpha}_n\big((id_E \otimes \kappa_F))_n(z); E, F''\big)$$

$$= \overleftarrow{\alpha}_n(z; E, F'') = \alpha_n(z; E, F'')$$

The same argument with E/K instead of E allows us to conclude $\alpha_n(z; E, F) \leq \alpha_n(z; E, F'')$. $\qquad\qquad\square$

We point out that part (b) above is a far reaching generalization of [1, Lem. 11] which only considers λ-o.s. tensor norms, the spaces are assumed to be C^*-algebras, and the embedding is only shown to be an isometry instead of a complete isometry.

Note that the completion of an operator space F has the same bidual as F. Therefore, as a consequence of the previous result we have:

Corollary 4.3.2 *Let α be an o.s. tensor norm (on* ONORM*) and let $\overline{F}$ be the completion of a normed operator space F. The mapping*

$$id_E \otimes \iota_F : E \otimes_\alpha F \to E \otimes_\alpha \overline{F},$$

where $\iota_F : F \to \overline{F}$ is the canonical inclusion, is a complete isometry with dense range in the following cases:

(a) α is finitely-generated and F is locally reflexive.
(b) α is an $\mathcal{E}(\lambda)$-o.s. tensor norm.
(c) α is cofinitely-generated.

4.4   The Density Lemma

We now state the so-called Density Lemma, which asserts that a given completely bounded mapping defined on the tensor product of certain normed operator spaces can be extended to a completely bounded mapping defined on the tensor product of the completions of the spaces. As usual, we need a local reflexivity hypothesis for the finitely-generated case. The proof easily follows from Corollary 4.3.2.

Lemma 4.4.1 *Let α be an o.s. tensor norm (on* ONORM*), E, F and G normed operator spaces, E_0 and F_0 dense subspaces of E and F respectively. Suppose $T \in CB(E \otimes_{\mathrm{proj}} F; G)$ such that $T|_{E_0 \otimes_\alpha F_0} \in CB(E_0 \otimes_\alpha F_0; G)$. Then $T \in CB(E \otimes_\alpha F; G)$ and*

$$\|T\|_{CB(E_0 \otimes_\alpha F_0; G)} = \|T\|_{CB(E \otimes_\alpha F; G)},$$

in the following cases:

(a) α is finitely-generated and F is locally reflexive.
(b) α is an $\mathcal{E}(\lambda)$-o.s. tensor norm.
(c) α is cofinitely-generated.

4.5   The $\mathcal{OS}_p$-local Technique Lemma

In the Banach space setting the $\mathscr{L}_p$-local Technique Lemma allows us to transfer results from tensor products of ℓ_p^n spaces to general $\mathscr{L}_p$-spaces (which, loosely speaking, are spaces whose finite dimensional subspaces "uniformly look" like ℓ_p^n's). In the operator space setting, one of the possible extensions of the $\mathscr{L}_p$-spaces are the $\mathcal{OS}_p$-spaces (see Sect. 1.1.1 for the definition). For this extension, we have the following local Technique Lemma.

Lemma 4.5.1 *Let α and β be o.s. tensor norms, $c \geq 0$, $1 \leq p \leq \infty$ and E a normed operator space such that for all $k \in \mathbb{N}$,*

$$\alpha \leq c\beta \quad on \quad E \otimes S_p^k.$$

Then $\alpha \leq Cc\,\vec{\beta} \quad on \quad E \otimes F$ for every $\mathcal{OS}_{p,C}$-space F.

Proof For $N \in \mathrm{OFIN}(F)$, by [46, Lem. 2.1] we can take a factorization

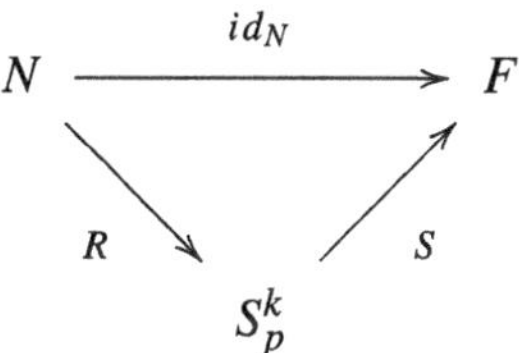

with $\|R\|_{\mathrm{cb}}\,\|S\|_{\mathrm{cb}} \leq C(1+\varepsilon)$. Let $M := S(S_p^k)$. Then for every $z \in M_n(E \otimes N)$,

$$\alpha_n(z; E, F) \leq \alpha_n((id_E \otimes (S \circ R))_n z; E, M) \leq \|S\|_{\mathrm{cb}}\,\alpha_n((id_E \otimes R)_n z; E, S_p^k)$$

$$\leq \|S\|_{\mathrm{cb}}\, c\beta_n((id_E \otimes R)_n z; F, S_p^k) \leq \|S\|_{\mathrm{cb}}\,\|R\|_{\mathrm{cb}}\, c\beta_n(z; F, N).$$

The statement follows. $\square$

Remark 4.5.2 Note that more generally the proof above applies whenever F has the γ_p-AP [46, Sec. 1], thanks to [46, Lem. 1.4]. Also note that the same argument yields versions of the local Technique Lemma for other operator space versions of the $\mathscr{L}_p$ spaces, e.g. the $\mathscr{OL}_p$ spaces from [46]. Furthermore, in the conclusion of Lemma 4.5.1 we can replace the finite hull by its right version (see Definition 10.3.3 below).

Chapter 5
Dual Operator Space Tensor Norms

We now turn our attention to understanding the dual of a tensor product equipped with an o.s. tensor norm. Ideally, one would like to describe it in terms of another tensor product involving the dual spaces. To establish this connection, we first need to examine the dual of an o.s. tensor norm, analyze its properties, and explore the results that follow from its structure. This will play a crucial role in later chapters, where tensor product techniques will be applied to the theory of mapping ideals.

As in the classical setting, we will obtain a version of the Duality Theorem in this context, along with a notion of accessibility. This theory will also be useful for some applications, including Chevet-Persson-Saphar-type inequalities and an analysis of the quantum Boolean cube. This demonstrates that tensor techniques serve as a valuable tool in their own right.

5.1 Duality Properties

In a nutshell, the dual o.s. tensor norm α' for the o.s. tensor norm α is the one that makes

$$E \otimes_{\alpha'} F = (E' \otimes_\alpha F')'$$

into a complete isometry for any $E, F \in$ OFIN. More explicitly, when $E, F \in$ OFIN for $z \in M_n(E \otimes F)$ we define

$$\alpha'_n(z; E, F) := \sup \left\{ \|\langle\langle u, z \rangle\rangle\|_{M_{mn}} \mid \alpha_m(u; E', F') \le 1 \right\}.$$

Let us check that, for example, the minimal and projective o.s. tensor norms are in duality with each other. Consider finite-dimensional operator spaces E and F. The operator space structure on $E \otimes_{\min} F$ is, by definition, the one induced

J. A. Chávez-Domínguez et al., *Operator Space Tensor Norms*, Lecture Notes in Mathematics 2379, https://doi.org/10.1007/978-3-031-96209-7_5

by the embedding $E \otimes_{\min} F \to \mathrm{CB}(E', F)$. Now, $(E' \otimes_{\mathrm{proj}} F')'$ is completely isometric to $\mathrm{CB}(E', F'') = \mathrm{CB}(E', F)$, from where we conclude that the canonical identification

$$E \otimes_{\min} F = (E' \otimes_{\mathrm{proj}} F')'$$

is a complete isometry. Taking duals and exchanging the roles of E, F and E', F' we also have a complete isometry

$$E \otimes_{\mathrm{proj}} F = (E' \otimes_{\min} F')'.$$

Proposition 5.1.1 *If α is an o.s. tensor norm on* OFIN, *then so is α'.*

Proof Let E, F be finite-dimensional operator spaces. It is clear that α' is an operator space structure on $E \otimes F$. From $\min \le \alpha \le \mathrm{proj}$, taking duals we conclude $\min = \mathrm{proj}' \le \alpha' \le \min' = \mathrm{proj}$ (for finite-dimensional spaces). The same sort of duality argument shows that α' is uniform. $\qquad\square$

The finite hull $\overrightarrow{\alpha'}$ of α' will be called the *dual o.s. tensor norm α'* (on ONORM) of the o.s. tensor norm α (on OFIN or ONORM).

The following properties are easy to check:

Proposition 5.1.2

(a) If $\alpha \le c\beta$, then $\beta' \le c\alpha'$.
(b) $\alpha = \alpha''$ on OFIN, *and $\overrightarrow{\alpha} = \alpha''$.*
(c) $\alpha = \alpha''$ on ONORM *if and only if α is finitely-generated.*

Since $\min$ and proj are finitely-generated, $\min' = \mathrm{proj}$ and $\mathrm{proj}' = \min$. Also, $h' = h$ (see, for instance, [34, Chap. 9]). Additionally, the symmetrized Haagerup o.s. tensor norms are in duality with each other.

Proposition 5.1.3 *If α and β are o.s. tensor norms,*

$$(\alpha \cap \beta)' = \alpha' + \beta' \quad and \quad (\alpha + \beta)' = \alpha' \cap \beta'.$$

In particular, $(h \cap h^t)' = h + h^t$ and $(h + h^t)' = h \cap h^t$.

Proof By Proposition 3.1.7 and Corollary 3.1.10 both $\alpha' \cap \beta'$ and $\alpha' + \beta'$ are finitely-generated, so it suffices to check that the dualities hold for finite-dimensional spaces. Let $E, F \in$ OFIN. By taking the adjoint of the canonical complete isometry $E \otimes_{\alpha \cap \beta} F \hookrightarrow (E \otimes_\alpha F) \oplus_\infty (E \otimes_\beta F)$ we get a canonical complete quotient $(E \otimes_\alpha F)' \oplus_1 (E \otimes_\beta F)' \twoheadrightarrow (E \otimes_{\alpha \cap \beta} F)'$. Since $(E \otimes_\alpha F)' = E' \otimes_{\alpha'} F'$ and $(E \otimes_\beta F)' = E' \otimes_{\beta'} F'$ we get a complete quotient $(E' \otimes_{\alpha'} F') \oplus_1 (E' \otimes_{\beta'} F') \twoheadrightarrow (E \otimes_{\alpha \cap \beta} F)'$. This is exactly the same quotient giving the o.s. structure to $E' \otimes_{\alpha' + \beta'} F'$, and thus $E' \otimes_{\alpha' + \beta'} F' = (E \otimes_{\alpha \cap \beta} F)'$ as claimed. The other equality follows analogously. $\qquad\square$

Definition 5.1.4 Given an o.s. tensor norm α, its *adjoint or contragradient* o.s. tensor norm is defined as $\alpha^* = (\alpha^t)' = (\alpha')^t$.

5.2   The Duality Theorem

As in the classical case, the following theorem tells us in what sense the completely isometric embedding

$$E \otimes_{\alpha'} F \to (E' \otimes_{\alpha} F')',$$

valid for finite-dimensional spaces, extends to infinite-dimensional ones.

Theorem 5.2.1 (The Duality Theorem)   *Let α be an o.s. tensor norm (on* OFIN*) and let E, F be normed operator spaces. The following natural mappings are complete isometries:*

$$E \otimes_{\overleftarrow{\alpha}} F \hookrightarrow (E' \otimes_{\alpha'} F')' \tag{5.2.1}$$

$$E' \otimes_{\overleftarrow{\alpha}} F \hookrightarrow (E \otimes_{\alpha'} F')' \qquad \textit{whenever } E \textit{ is locally reflexive} \tag{5.2.2}$$

$$E' \otimes_{\overleftarrow{\alpha}} F' \hookrightarrow (E \otimes_{\alpha'} F)' \qquad \textit{whenever } E \textit{ and } F \textit{ are locally reflexive} \tag{5.2.3}$$

Proof   To prove (5.2.1), observe first that

$$\mathrm{OFIN}(E') = \{K^0 \mid K \in \mathrm{OCOFIN}(E)\}$$

and that for $(K, L) \in \mathrm{OCOFIN}(E) \times \mathrm{OCOFIN}(F)$, $z \in M_n(E \otimes F)$ and $u \in M_m(K^0 \otimes L^0) \subset M_m(E' \otimes F')$, we have

$$\langle\langle z, u \rangle\rangle = \langle\langle (q_K^E \otimes q_L^F)(z), u \rangle\rangle$$

Now, by the valid duality relation for finite-dimensional spaces,

$$\overleftarrow{\alpha}_n(z; E, F) = \sup_{K,L} \left\{ \alpha_n\big((q_K^E \otimes q_L^F)(z); E/K, F/L\big) \right\}$$

$$= \sup_{K,L} \sup \left\{ \left\| \langle\langle (q_K^E \otimes q_L^F)(z), u \rangle\rangle \right\|_{M_{mn}} \mid \alpha'_m(u; K^0, L^0) < 1 \right\}$$

$$= \sup \left\{ \| \langle\langle z, u \rangle\rangle \|_{M_{mn}} \mid \overrightarrow{\alpha'}_m(u; E', F') < 1 \right\}$$

and this is precisely (5.2.1).

Consider now the commutative diagram

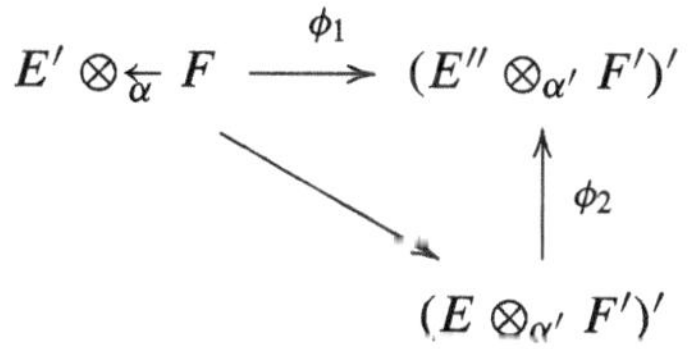

$$
\begin{array}{ccc}
E' \otimes_{\overleftarrow{\alpha}} F & \xrightarrow{\ \phi_1\ } & (E'' \otimes_{\alpha'} F')' \\
 & \searrow & \big\uparrow{\scriptstyle \phi_2} \\
 & & (E \otimes_{\alpha'} F')'
\end{array}
$$

Note that ϕ_1 is a complete isometry by (5.2.1), whereas ϕ_2 is a complete isometry when E is locally reflexive by the Extension Lemma 4.2.1. Thus, (5.2.2) follows. In the same way, in the diagram

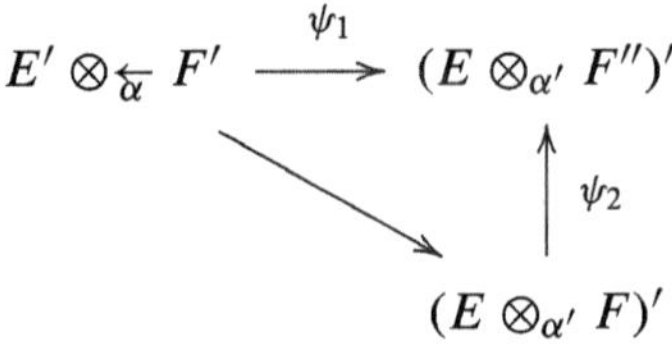

ψ_1 is a complete isometry (if E is locally reflexive) by (5.2.2) and ψ_2 is a complete isometry (if F is locally reflexive) due to the Extension Lemma 4.2.1. This proves (5.2.3). $\qquad\square$

Remark 5.2.2 The Extension Lemma 4.2.1 is valid without the hypothesis of local reflexivity for $\mathcal{E}(\lambda)$-o.s. tensor norms. Hence, in Eqs. (5.2.2) and (5.2.3) of the Duality Theorem, if α' is an $\mathcal{E}(\lambda)$-o.s. tensor norm the local reflexivity of the involved spaces is not needed.

In particular, for $\alpha' = \text{proj}$, recalling that $\overleftarrow{\min} = \min$ we obtain for arbitrary normed operator spaces E and F the following known completely isometric embeddings

$$E' \otimes_{\min} F \hookrightarrow (E \otimes_{\text{proj}} F')' \quad \text{and} \quad E' \otimes_{\min} F' \hookrightarrow (E \otimes_{\text{proj}} F)'.$$

One nice consequence of the Duality Theorem is the following:

Corollary 5.2.3 *If E and F are normed operator spaces with the completely metric approximation property, then the natural mappings*

$$E \otimes_\alpha F \to E \otimes_{\overleftarrow{\alpha}} F \quad \text{and} \quad E \otimes_\alpha F \to (E' \otimes_{\alpha'} F')'$$

are complete isometries.

Proof The complete isometry between $E \otimes_\alpha F$ and $E \otimes_{\overleftarrow{\alpha}} F$ was already shown in Corollary 4.1.2. Then, (5.2.1) implies the rest. $\qquad\square$

Observe that if we replace the CMAP property by the CBAP property the above identifications become complete isomorphisms (with norms controlled by the CBAP constant).

We now show that the Haagerup o.s. tensor norm is both finitely and cofinitely-generated.

Remark 5.2.4 Let h be the Haagerup o.s. tensor norm. Then,

$$h = \overleftarrow{h} = \overrightarrow{h}.$$

Proof The fact that $h = \overrightarrow{h}$ was already mentioned after Definition 3.1.1.

To see that $h = \overleftarrow{h}$ notice that by the Duality Theorem 5.2.1 the canonical embedding below is a complete isometry

$$E \otimes_{\overleftarrow{h}} F \hookrightarrow (E' \otimes_{h'} F')' = (E' \otimes_h F')', \tag{5.2.4}$$

where we have used in the last equality the self-dual property of the o.s. tensor norm h [34, Cor. 9.4.8]. On the other hand, by [34, Thm. 9.4.7.] and the Embedding Lemma 4.3.1 we have that

$$E \otimes_h F \hookrightarrow E'' \otimes_h F'' \hookrightarrow (E' \otimes_h F')', \tag{5.2.5}$$

is a complete isometry. Since both complete isometries in Eqs. (5.2.4) and (5.2.5) are the same canonical inclusion, h and $\overleftarrow{h}$ must coincide as o.s. tensor norms.   □

5.3   Accessible Operator Space Tensor Norms

In general, operator space tensor norms are not both finitely and cofinitely generated; in other words, the equality $\overrightarrow{\alpha} = \overleftarrow{\alpha}$ does not always hold on ONORM. However, for many purposes, it is often sufficient that the equality is valid on finite-dimensional spaces or on locally reflexive spaces. This motivates the next definitions, which we will use in the sequel.

Definition 5.3.1 An o.s. tensor norm α is called *right-accessible* if

$$\overrightarrow{\alpha}_n(\cdot; M, F) = \overleftarrow{\alpha}_n(\cdot; M, F) \qquad \text{for all } M \in \text{OFIN}, F \in \text{ONORM}, n \in \mathbb{N}$$

and *left-accessible* if

$$\overrightarrow{\alpha}_n(\cdot; E, M) = \overleftarrow{\alpha}_n(\cdot; E, M) \qquad \text{for all } M \in \text{OFIN}, E \in \text{ONORM}, n \in \mathbb{N}.$$

An o.s. tensor norm that is both left and right-accessible is called *accessible*, and *totally accessible* means that $\overrightarrow{\alpha} = \overleftarrow{\alpha}$.

In many cases, local reflexivity naturally appears when dealing with accessibility. Based on this we introduce the following:

Definition 5.3.2 An o.s. tensor norm α is called *locally right-accessible if*

$$\overrightarrow{\alpha}_n(\cdot; M, F) = \overleftarrow{\alpha}_n(\cdot; M, F) \qquad \text{for all } M \in \text{OFIN}, F \in \text{OLOC}, n \in \mathbb{N}$$

and *locally left-accessible* if

$$\overrightarrow{\alpha}_n(\cdot; E, M) = \overleftarrow{\alpha}_n(\cdot; E, M) \qquad \text{for all } M \in \text{OFIN}, E \in \text{OLOC}, n \in \mathbb{N}$$

An o.s. tensor norm that is both locally left and locally right-accessible is called *locally accessible*, and *locally totally accessible* means that $\overrightarrow{\alpha} = \overleftarrow{\alpha}$ in $\mathrm{OLOC} \otimes \mathrm{OLOC}$.

We have already seen that the minimal and the Haagerup o.s. tensor norms are both finitely and cofinitely-generated, or in other words they are totally accessible. Recall that, as we saw in Proposition 3.1.3, the projective o.s. tensor norm is not cofinitely-generated, hence it is not totally accessible. Nevertheless, as in the Banach space framework, proj is indeed accessible. To see this, let $E \in \mathrm{ONORM}$ and $M \in \mathrm{OFIN}$. We always have that the natural mapping

$$E \otimes_{\mathrm{proj}} M \to (E' \otimes_{\mathrm{min}} M')' \tag{5.3.1}$$

is a complete isometry (see [31, Thm. 2.2] or Remark 8.2.8). On the other hand, as a consequence of the Duality Theorem 5.2.1 we also have that the mapping

$$E \otimes_{\underset{\mathrm{proj}}{\longleftarrow}} M \to (E' \otimes_{\mathrm{min}} M')'$$

is a complete isometry. Thus, both o.s. tensor norms, $\overrightarrow{\mathrm{proj}} = \mathrm{proj}$ and $\overleftarrow{\mathrm{proj}}$, must coincide on $E \otimes M$. This shows the left-accessibility of proj. Analogously, we have that proj is also right-accessible and therefore accessible.

More generally, a similar argument will show that any λ-o.s. tensor norm is accessible. Let us start with a formula for calculating the dual of a λ-o.s. tensor norm, which is essentially [75, Prop. 5.3].

Proposition 5.3.3 *Let E and F be operator spaces, and $u \in M_n(E \otimes F)$. For any λ-o.s. tensor norm,*

$$\lambda'_n(u; E, F) = \sup \left\{ \left\| \left(\otimes_{B_k^\lambda} (\phi, \psi) \right)_n (u) \right\|_{M_{n\tau(k)}} : \right.$$
$$\left. k \in \mathbb{N}, \ \|\phi\|_{M_k(E')}, \ \|\psi\|_{M_k(F')} \le 1 \right\}$$

Proof Assume first that E and F are finite-dimensional. By the definition of λ' and Theorem 2.2.10,

$$E \otimes_{\lambda'} F = (E' \otimes_\lambda F')' = \mathrm{CB}_\lambda(E' \times F').$$

Therefore, $\lambda'_n(u; E, F)$ is the norm $\|\Phi_u\|_{\mathrm{cb},\lambda}$ of the bilinear map $\Phi_u : E' \times F' \to M_n$ associated to u. By definition, said norm is

$$\|\Phi_u\|_{\mathrm{cb},\lambda} = \sup_{k \in \mathbb{N}} \left\{ \left\| (\Phi_u)_{B_k^\lambda}(\phi, \psi) \right\|_{M_{n\tau(k)}} : \|\phi\|_{M_k(E')}, \ \|\psi\|_{M_k(F')} \le 1 \right\},$$

so it suffices to check

$$\left(\otimes_{B_k^\lambda}(\phi,\psi)\right)_n(u) = (\Phi_u)_{B_k^\lambda}(\phi,\psi)$$

which follows easily from the definitions.

Let us now consider the case of arbitrary E and F. Let $E_0 \in \mathrm{OFIN}(E)$ and $F_0 \in \mathrm{OFIN}(F)$ such that $u \in M_n(E_0 \otimes F_0)$. By the previous argument,

$$\lambda'_n(u; E_0, F_0) = \sup\left\{ \left\|\left(\otimes_{B_k^\lambda}(\phi_0,\psi_0)\right)_n(u)\right\|_{M_{n\tau(k)}} : \right.$$
$$\left. k \in \mathbb{N}, \|\phi_0\|_{M_k(E_0')}, \|\psi_0\|_{M_k(F_0')} \leq 1 \right\}.$$

By the Arveson extension theorem, $\phi_0 \in M_k(E_0') = \mathrm{CB}(E_0, M_k)$ and $\psi_0 \in M_k(F_0') = \mathrm{CB}(F_0, M_k)$ admit respective norm-preserving extensions $\phi \in \mathrm{CB}(E, M_k)$ and $\psi \in \mathrm{CB}(F, M_k)$. Since clearly

$$\left(\otimes_{B_k^\lambda}(\phi,\psi)\right)_n(u) = \left(\otimes_{B_k^\lambda}(\phi_0,\psi_0)\right)_n(u),$$

it follows that

$$\lambda'_n(u; E_0, F_0) = \sup\left\{ \left\|\left(\otimes_{B_k^\lambda}(\phi,\psi)\right)_n(u)\right\|_{M_{n\tau(k)}} : \right.$$
$$\left. k \in \mathbb{N}, \|\phi\|_{M_k(E')}, \|\psi\|_{M_k(F')} \leq 1 \right\},$$

and therefore $\lambda'_n(u; E, F)$ is also equal to the same quantity. $\qquad\square$

As an immediate consequence, we have the following result (which in [75, Def. 5.1] is taken as the definition for the dual of λ).

Corollary 5.3.4 *Let E and F be operator spaces. For any λ-o.s. tensor norm, the natural map*

$$E \otimes_{\lambda'} F \hookrightarrow \mathrm{CB}_\lambda(E' \times F')$$

is a complete isometry.

Alternatively, we have another outcome.

Corollary 5.3.5 *Let E and F be operator spaces. For any λ-o.s. tensor norm, the natural mapping*

$$E' \otimes_{\lambda'} F' \hookrightarrow \mathrm{CB}_\lambda(E \times F)$$

is a complete isometry.

Proof From the previous result, the natural map

$$E' \otimes_{\lambda'} F' \hookrightarrow \mathrm{CB}_\lambda(E'' \times F'') \tag{5.3.2}$$

is a complete isometry. By the Extension Lemma 4.2.5 we also have a canonical completely isometric embedding

$$\mathrm{CB}_\lambda(E \times F) \hookrightarrow \mathrm{CB}_\lambda(E'' \times F''). \qquad (5.3.3)$$

It is clear that the image of the map in (5.3.2) is contained in that of the map in (5.3.3), which gives the desired result. $\qquad\square$

In the case where E is finite-dimensional the complete isometry from the previous proposition gives rise to the next completely isometric mapping.

Proposition 5.3.6 *Let E and F be operator spaces, with E finite-dimensional. For any λ-o.s. tensor norm, the natural map*

$$E \otimes_\lambda F \hookrightarrow (E' \otimes_{\lambda'} F')'$$

is a complete isometry.

Proof Since E is finite-dimensional, the complete isometry given in the previous corollary

$$E' \otimes_{\lambda'} F' \hookrightarrow \mathrm{CB}_\lambda(E'' \times F'')$$

is surjective. Therefore, $E' \otimes_{\lambda'} F' = \mathrm{CB}_\lambda(E'' \times F'')$. Taking duals and using Theorem 2.2.10 we have

$$(E \otimes_\lambda F'')'' = \mathrm{CB}_\lambda(E'' \times F'')' = (E' \otimes_{\lambda'} F')',$$

which, together with the Embedding Lemma 4.3.1, finishes the proof. $\qquad\square$

Now we reach the promised argument about the accessibility of λ-o.s. tensor norms.

Theorem 5.3.7 *Any λ-o.s. tensor norm is accessible.*

Proof Let E and F be operator spaces, with E finite-dimensional. By Proposition 5.3.6 and the Duality Theorem 5.2.1, both mappings

$$E \otimes_\lambda F \hookrightarrow (E' \otimes_{\lambda'} F')' \quad \text{and} \quad E \otimes_{\overleftarrow{\lambda}} F \hookrightarrow (E' \otimes_{\lambda'} F')'$$

are complete isometries, and therefore $E \otimes_\lambda F = E \otimes_{\overleftarrow{\lambda}} F$. Since any λ-o.s. tensor norm is finitely-generated we conclude that λ is right-accessible. The proof for left-accessibility is analogous. $\qquad\square$

The following lemma is the key to proving that the intersection of two cofinitely-generated o.s. tensor norms is cofinitely-generated.

Lemma 5.3.8 *Let E and F be operator spaces. Suppose that for the o.s. tensor norms α and β, the natural maps*

$$E \otimes_\alpha F \hookrightarrow (E' \otimes_{\alpha'} F')', \qquad E \otimes_\beta F \hookrightarrow (E' \otimes_{\beta'} F')'$$

are complete isometries. Then so is the natural map

$$E \otimes_{\alpha \cap \beta} F \hookrightarrow (E' \otimes_{(\alpha \cap \beta)'} F')'.$$

Proof We have natural completely isometric inclusions

$$E \otimes_{\alpha \cap \beta} F \hookrightarrow (E' \otimes_{\alpha'} F')' \oplus_\infty (E' \otimes_{\beta'} F')'$$

$$(E' \otimes_{\alpha'+\beta'} F')' \hookrightarrow \left[(E' \otimes_{\alpha'} F') \oplus_1 (E' \otimes_{\beta'} F') \right]'.$$

Since the spaces on the right are canonically identified, it suffices to observe that the image of the first map is contained in the image of the second one and apply Proposition 5.1.3. $\qquad\qquad\square$

Corollary 5.3.9 *The o.s. tensor norm $h \cap h^t$ is totally accessible.*

Proof Let E and F be operator spaces. Recall that h and h^t are totally accessible. Then the previous Lemma and Proposition 5.1.3 yield that the natural map

$$E \otimes_{h \cap h^t} F \hookrightarrow (E' \otimes_{h+h^t} F')'$$

is a complete isometry. Now, by the Duality Theorem 5.2.1 there is also completely isometric the mapping

$$E \otimes_{\overleftarrow{h \cap h^t}} F \hookrightarrow (E' \otimes_{h+h^t} F')'.$$

This means that $E \otimes_{h \cap h^t} F = E \otimes_{\overleftarrow{h \cap h^t}} F$ and taking into account that $h \cap h^t$ is finitely-generated (see Proposition 3.1.6) we obtain that $h \cap h^t$ is totally accessible.
$\qquad\qquad\square$

Remark 5.3.10 Arguing as in the proof of the previous corollary we have: let α and β be cofinitely-generated o.s. tensor norms, then $\alpha \cap \beta$ is also cofinitely-generated.

Left-accessibility of an o.s. tensor norm α provides a complete isometry analogous to the one that appears (for the o.s. tensor norm proj) in Eq. (5.3.1). This fact combined with a CMAP hypothesis imply the following result.

Corollary 5.3.11

(a) *Let $E \in$ ONORM, F a normed operator space with the CMAP and α a left-accessible o.s. tensor norm. Then,*

$$E \otimes_{\overrightarrow{\alpha}} F \to E \otimes_{\overleftarrow{\alpha}} F \qquad and \qquad E \otimes_{\overrightarrow{\alpha}} F \to (E' \otimes_{\alpha'} F')'$$

are complete isometries.

(b) Let $E \in$ OLOC, F a normed operator space with the CMAP and α a locally left-accessible o.s. tensor norm. Then,

$$E \otimes_{\overrightarrow{\alpha}} F \to E \otimes_{\overleftarrow{\alpha}} F \qquad and \qquad E \otimes_{\overrightarrow{\alpha}} F \to (E' \otimes_{\alpha'} F')'$$

are complete isometries.

(c) The analogous results hold with right-accessibility/local right-accessibility.

Proof We will only prove (a) since the proofs of the other statements are similar. Note that, by the left-accessibility of the o.s. tensor norm we have

$$\overrightarrow{\alpha} = \overleftarrow{\alpha} \qquad on \qquad E \otimes M \tag{5.3.4}$$

for every $M \in$ OFIN(F). Now by the Approximation Lemma 4.1.1, since F has the CMAP then $\overrightarrow{\alpha}$ and $\overleftarrow{\alpha}$ coincide on $E \otimes F$, and therefore

$$E \otimes_{\overrightarrow{\alpha}} F \to E \otimes_{\overleftarrow{\alpha}} F$$

is a complete isometry. The fact that $E \otimes_{\overrightarrow{\alpha}} F \to (E' \otimes_{\alpha'} F')'$ is also a complete isometry follows from the previous identification and the Duality Theorem 5.2.1.

$\square$

Note that by Proposition 3.1.2 in the statement of the previous corollary we can change $\overrightarrow{\alpha}$ by α and, of course, everything remains the same. Let us now consider the issue of the accessibility of dual o.s. tensor norms.

Proposition 5.3.12 *Let α be an o.s. tensor norm on* ONORM.

(a) If α is right-accessible (resp. left-accessible, accessible) then α' is locally right-accessible (resp. locally left-accessible, locally accessible).

(b) If α is accessible then the transposed o.s. tensor norm α^t is accessible and the adjoint o.s. tensor norm α^* is locally accessible.

Proof Assume that α is right-accessible. Then for all $M \in$ OFIN and $F \in$ OLOC we have complete isometries

$$M' \otimes_{\alpha''} F' = M' \otimes_{\overrightarrow{\alpha}} F' = M' \otimes_{\overleftarrow{\alpha}} F' = (M \otimes_{\alpha'} F)'$$

where the first equality follows from Proposition 5.1.2, the second one from right-accessibility, and the third one from the Duality Theorem 5.2.1. Therefore we also have complete isometries

$$M \otimes_{\alpha'} F \hookrightarrow (M \otimes_{\alpha'} F)'' = (M' \otimes_{\alpha''} F')',$$

which means that $\overleftarrow{\alpha'} = \alpha' = \overrightarrow{\alpha'}$ (again by the Duality Theorem) in OFIN $\otimes$ OLOC. This shows that α' is locally right-accessible. Part (b) follows trivially. $\square$

In the Banach space setting the dual of a right(left)-accessible tensor norm is again right(left)-accessible. For operator spaces we obtain in the previous proposition a weaker statement since we have proved that the dual norm is *locally* right(left)-accessible. However, we do not know whether this notion is truly weaker since we do not have any example of an o.s. tensor norm which is locally right(left)-accessible but not right(left)-accessible.

As usual, when dealing with $\mathcal{E}(\lambda)$-o.s. tensor norms we can avoid the local reflexivity conditions. Thus, the corresponding version of Proposition 5.3.12 runs as follows:

Corollary 5.3.13 *Let α be an o.s. tensor norm on* ONORM *and suppose that α' is an $\mathcal{E}(\lambda)$-o.s. tensor norm.*

(a) If α is right-accessible (resp. left-accessible, accessible) then α' is right-accessible (resp. left-accessible, accessible).

(b) If α is accessible then the transposed o.s. tensor norm α^t is accessible and the adjoint o.s. tensor norm α^* is accessible.

In particular, due to Corollary 5.3.9, the o.s. tensor norm $h + h^t$ is accessible.

Remark 5.3.14 Recall that total accessibility does not transfer from a norm to its dual norm, as can be seen by noting that min is totally accessible while proj is not. We do not know whether the o.s. tensor norm $h + h^t$ is totally accessible as its dual norm $h \cap h^t$. In any case, it would not be surprising if it is not. Given that $h + h^t$ is finitely-generated (Proposition 3.1.6), we can reformulate this question by asking if $h + h^t$ is cofinitely-generated.

5.4   Chevet-Persson-Saphar Inequalities in the Noncommutative Setting

For a measure μ and a Banach space X, one can consider the "natural" norm on $L_p(\mu) \otimes X$ induced by its inclusion in $L_p(\mu; X)$. This is not "tensorial" in the sense that it does not satisfy multiplicativity: given bounded linear maps $T : L_p(\mu) \to L_p(\nu)$ and $S : X \to Y$, we do not necessarily have that the norm of $T \otimes S : L_p(\mu; X) \to L_p(\nu; Y)$ is bounded by $\|T\| \, \|S\|$. For example, if $T : \ell_2(\mathbb{Z}) \to \ell_2(\mathbb{Z})$ is the discrete Hilbert transform, and $S = Id_{\ell_1} : \ell_1 \to \ell_1$ is the identity on ℓ_1, then $T \otimes S$ is not even bounded [23, 7.6]. While this failure of multiplicativity seems disappointing at first sight, it turns out to have highly interesting consequences. See [23, Chap. 7] for more details.

Similarly, for a Hilbert space H and an operator space E we can consider the "natural" operator space structure on $S_p(H) \otimes E$ induced from $S_p[H; E]$. We will denote this by $S_p(H) \otimes_{\Delta_p} E$, a notation that will allow for the convenient expression "$\alpha < c\Delta_p$ on $S_p(H) \otimes E$" to mean that the identity map $S_p(H) \otimes_{\Delta_p} E \to S_p(H) \otimes_\alpha E$ has cb-norm less than or equal to c (and we define $\Delta_p \leq c\alpha$

analogously). We also write $E \otimes_{\Delta_p^t} S_p(H)$ for $E \otimes S_p(H)$ with the operator space structure induced by the flip $E \otimes_{\Delta_p^t} S_p(H) = S_p(H) \otimes_{\Delta_p} E$.

It is immediate that $\otimes_{\Delta_p}$ will generally not satisfy multiplicativity, using the same example of the discrete Hilbert transform above by realizing $\ell_2(\mathbb{Z})$ inside S_2 as diagonal operators (note that the resulting operator $T : S_2 \to S_2$ is bounded and hence completely bounded by the homogeneity of S_2, see [65, Prop. 7.2.(iii) and Rmk. (iv) p.129]).

However, just as in the classical case, the "natural" operator space structure on $S_p(H) \otimes H$ is confined between two well-established operator space tensor norms. In the classical setting this is expressed by the Chevet-Persson-Saphar inequalities [21, 55, 72]. Here we will prove an operator space version of them. We point out that a closely related result appeared in [18, Thm. 4.2].

We start with some technical lemmas. The first one is the essential ingredient to show that $d_p \leq \Delta_p$.

Lemma 5.4.1 *Let* $1 \leq p \leq \infty$ *and let* E *be an operator space. The formal identity* $S_p^n[E] \to S_p^n \otimes_{d_p} E$ *is completely contractive.*

Proof Define a map $\varphi : S_p^n[E] \to (S_{p'}^n \otimes_{\min} S_p^n) \otimes_{\text{proj}} S_p^n[E]$ by $x \mapsto u \otimes x$ where $u \in S_{p'}^n \otimes_{\min} S_p^n$ is the element corresponding to the identity map $S_{p'}^n \to S_{p'}^n$. Note that φ is completely contractive since proj is an o.s. tensor norm. An easy calculation shows that the diagram

$$
\begin{array}{ccc}
 & (S_{p'}^n \otimes_{\min} S_p^n) \otimes_{\text{proj}} S_p^n[E] & \\
{\scriptstyle \varphi}\nearrow & & \downarrow {\scriptstyle q^{d_p,n}} \\
S_p^n[E] \xrightarrow{\hspace{2cm}} & S_p^n \otimes_{d_p^n} E &
\end{array}
$$

is commutative, showing that $S_p^n[E] \to S_p^n \otimes_{d_p^n} E$ is completely contractive which implies the desired result since $d_p \leq d_p^n$. $\square$

The second lemma will show how d_p and the dual norm of $g_{p'}$ are related in OFIN.

Lemma 5.4.2 *For any* $E, F \in$ OFIN *and any* $1 \leq p \leq \infty$, *the natural map*

$$E \otimes_{d_p} F \to (E' \otimes_{g_{p'}} F')' = E \otimes_{g_{p'}'} F$$

is completely contractive.

Proof We will prove that for every $k, m \in \mathbb{N}$ the natural map

$$E \otimes_{d_p^k} F \to (E' \otimes_{g_{p'}^m} F')'$$

is completely contractive, and the desired result will follow by taking limits (or by repeating the same arguments using the tensor contraction q^{d_p} from Proposition 2.2.4). By the identification

$$\mathrm{CB}\left(E \otimes_{d_p^k} F, (E' \otimes_{g_{p'}^m} F')'\right) = \left((E \otimes_{d_p^k} F) \otimes_{\mathrm{proj}} (E' \otimes_{g_{p'}^m} F')\right)'$$

the corresponding map (i.e. the tensor contraction)

$$(E \otimes_{d_p^k} F) \otimes_{\mathrm{proj}} (E' \otimes_{g_{p'}^m} F') \to \mathbb{C}$$

has the same norm. By the projectivity of proj and the definition of d_p^k and $g_{p'}^m$ in terms of complete quotients, we can consider instead the norm of the tensor contraction

$$\left((S_{p'}^k \otimes_{\mathrm{min}} E) \otimes_{\mathrm{proj}} S_p^k[F]\right) \otimes_{\mathrm{proj}} \left(S_{p'}^m[E'] \otimes_{\mathrm{proj}} (S_p^m \otimes_{\mathrm{min}} F')\right) \to \mathbb{C}. \quad (5.4.1)$$

Since the projective tensor product is commutative, it will suffice to show that the tensor contractions

$$(S_{p'}^k \otimes_{\mathrm{min}} E) \otimes_{\mathrm{proj}} S_{p'}^m[E'] \to S_{p'}^m[S_{p'}^k], \qquad (S_p^m \otimes_{\mathrm{min}} F') \otimes_{\mathrm{proj}} S_p^k[F] \to S_p^k[S_p^m]$$
$$(5.4.2)$$

are completely contractive, because then the map in (5.4.1) will be completely contractive by the duality between $S_{p'}^k[S_{p'}^m]$ and $S_p^m[S_p^k]$. Let us see the argument for the first map in (5.4.2); the proof for the other one is clearly analogous. First, by the identification

$$\mathrm{CB}((S_{p'}^k \otimes_{\mathrm{min}} E) \otimes_{\mathrm{proj}} S_{p'}^m[E'], S_{p'}^m[S_{p'}^k]) = \mathrm{CB}\left(S_{p'}^k \otimes_{\mathrm{min}} E, \mathrm{CB}(S_{p'}^m[E'], S_{p'}^m[S_{p'}^k])\right)$$

we can instead consider the norm of the corresponding element in the latter space. But this is completely contractive because is the composition of the completely isometric inclusions

$$S_{p'}^k \otimes_{\mathrm{min}} E \hookrightarrow \mathrm{CB}(E', S_{p'}^k) \hookrightarrow \mathrm{CB}\left(S_{p'}^m[E'], S_{p'}^m[S_{p'}^k]\right),$$

where the first one follows from the definition of the minimal tensor product, and the second one from Theorem 1.4.3.   □

Recall from Theorem 2.2.8 that on any tensor product $E \otimes F$ the o.s. tensor norm d_1 coincides with proj, while d_∞ has a more complicated relationship to min. We have already observed that, as a consequence of Theorem 2.2.8, d_∞ and min coincide when the space F is of the form $S_1(H)$. We now show that d_∞ also coincides with min when the space F is of the form $S_\infty(H)$.

Corollary 5.4.3 *For any Hilbert space H and any operator space F, $d_\infty = \min$ on $S_\infty(H) \otimes F$.*

Proof We always have $\min \leq d_\infty$, and $d_\infty \leq \min$ on $S_\infty^n \otimes F$ follows from Lemma 5.4.1 since $S_\infty^n[F] = S_\infty^n \otimes_{\min} F$. Since $S_\infty(H)$ is a $\mathcal{OS}_{\infty,1+\varepsilon}$-space for every $\varepsilon > 0$ and $\min$ is finitely-generated, from the Local Technique Lemma 4.5.1 we have then $d_\infty \leq \min$ on $S_\infty(H) \otimes F$. $\square$

We are now ready to prove the main result of this section, the Chevet-Persson-Saphar inequalities for operator spaces.

Theorem 5.4.4 *For any Hilbert space H, any operator space E and any $1 \leq p \leq \infty$, we have*

$$d_{p'}^* \leq d_p \leq \Delta_p \leq g_{p'}^* \leq g_p \qquad \text{on } S_p(H) \otimes E.$$

Proof Since $g_{p'}'$ and d_p are both finitely-generated (the former because all dual o.s. tensor norms are finitely-generated by definition, and the latter by Proposition 3.1.5), to verify $d_{p'}^* = g_{p'}' \leq d_p$ it suffices to check it on finite-dimensional spaces. This is precisely Lemma 5.4.2. The inequality $g_{p'}^* \leq g_p$ follows by transposing, so now we have proved the leftmost and rightmost inequalities in the statement.

For any finite-dimensional $K \subseteq H$, from Lemma 5.4.1 we have that the identity $S_p(K) \otimes_{\Delta_p} E \to S_p(K) \otimes_{d_p} E$ is completely contractive. The uniformity of d_p gives the complete contractivity of $S_p(K) \otimes_{d_p} E \to S_p(H) \otimes_{d_p} E$, since $S_p(K) \subseteq S_p(H)$ completely isometrically. We thus have $S_p(K) \otimes_{\Delta_p} E \to S_p(H) \otimes_{d_p} E$ is completely contractive. By considering the union of all the subpaces $S_p(K) \otimes_{\Delta_p} E \subset S_p(H) \otimes_{\Delta_p} E$ as K ranges over all finite-dimensional subspaces of H, we then have $d_p \leq \Delta_p$ on a dense subspace of $S_p(H) \otimes_{\Delta_p} E$ (which implies it on all of $S_p(H) \otimes_{\Delta_p} E$).

For the remaining inequality $\Delta_p \leq g_{p'}^*$, we consider $p = \infty$ first. In this case, we have on $S_\infty(H) \otimes E$ that $\Delta_\infty = \min$, and clearly $\min \leq g_1^*$ since the latter is an o.s. tensor norm. For $1 \leq p < \infty$, consider the diagram of natural maps

$$
\begin{array}{ccc}
S_p(H) \otimes_{\Delta_p} E & \longrightarrow & \left(S_{p'}(H) \otimes_{\Delta_{p'}} E'\right)' \\
& & \uparrow \\
\\
S_p(H) \otimes_{\underset{d_{p'}'}{\longleftarrow}} E & \longrightarrow & \left(S_{p'}(H) \otimes_{d_{p'}} E'\right)'
\end{array}
$$

The vertical arrow is a complete contraction by what has already been shown. The bottom horizontal arrow is a complete isometry by the Duality Theorem 5.2.1. The top horizontal arrow is also a complete isometry, since so is the canonical inclusion $E \to E''$ and therefore so is $S_p(H) \otimes_{\Delta_p} E \to S_p(H) \otimes_{\Delta_p} E''$ (Theorem 1.4.3), plus the identification $\left(S_{p'}(H) \otimes_{\Delta_{p'}} E'\right)' = S_p(H)\widehat{\otimes}_{\Delta_p} E''$ (Theorem 1.4.4, here is where we need $p' \neq 1$ or $p \neq \infty$). We have then shown $\Delta_p \leq \overleftarrow{d'_{p'}} \leq d'_{p'} = g^*_{p'}$. $\square$

As a very useful consequence, we can now prove that the o.s. tensor norms d_p and g_p implement the Fubini theorem for (scalar-valued) S_p spaces. This was first proved in [18, Cor. 5.1].

Corollary 5.4.5 *For any Hilbert spaces H and K, and $1 \leq p \leq \infty$, $S_p(H)\widehat{\otimes}_{d_p} S_p(K)$ is completely isometric to $S_p(H \otimes_2 K)$.*

Proof From the Chevet-Persson-Saphar inequalities (Theorem 5.4.4) we have the following commutative diagram of complete contractions

$$
\begin{array}{ccccc}
S_p(H)\widehat{\otimes}_{g_p} S_p(K) & \longrightarrow & S_p[H; S_p(K)] & \longrightarrow & S_p(H)\widehat{\otimes}_{d_p} S_p(K) \\
\uparrow & & & & \downarrow \\
S_p(K)\widehat{\otimes}_{d_p} S_p(H) & \longleftarrow & S_p[K; S_p(H)] & \longleftarrow & S_p(K)\widehat{\otimes}_{g_p} S_p(H)
\end{array}
$$

where the vertical arrows are the complete isometries given by flipping. Note that all the maps involved are canonical, either the identity or a flip, so all of the arrows are in fact complete isometries. By the Fubini theorem for vector valued S_p spaces (Theorem 1.4.5), we conclude that all of the spaces in the diagram above are (canonically) completely isometric to $S_p(H \otimes_2 K)$. $\square$

5.5   Connections Between o.s. Tensor Norms and Quantum Analysis

The classical operator space tensor products (min, proj, $h, h + h^t$) have a well-deserved place in the theory of operator spaces due to their applications in Operator Algebras and beyond. The reader might justifiably wonder whether there is much of a point in developing operator space versions of other more obscure parts of the theory of tensor products of Banach spaces. To whet the reader's appetite we illustrate the potential for such applications with an example, showing how some of the more abstract operator space tensor products we have developed have fairly immediate consequences in areas such as analysis on the quantum Boolean cube and Quantum Information Theory. The material in this section will not be used later on, so it can be read independently at any time. While some prior knowledge in Quantum Analysis may be necessary to fully appreciate its significance, we emphasize the relevance of the applications presented, which provide motivation for the theoretical developments.

First, let us recall the case of the classical Boolean cube. Our presentation borrows heavily from [11, 43]. Given $n \in \mathbb{N}$, let $\{-1, 1\}^n$ be the *Boolean (or*

Hamming) cube of dimension n, that is, the set of vectors $x = (x_1, \ldots, x_n)$ such that $x_j = \pm 1$ for all $j = 1, \ldots, n$. When we write $L_p(\{-1, 1\}^n)$ the measure under consideration is the uniform measure on $\{-1, 1\}^n$, and we denote the norm in this space by $\|\cdot\|_p$. For a subset $S \subseteq [n] := \{1, 2, \ldots, n\}$, the *Walsh function* $\chi_S : \{-1, 1\}^n \to \mathbb{R}$ is defined as $\chi_S(x) := \prod_{j \in S} x_j$ and $\chi_\emptyset(x) := 1$. The system $\{\chi_S : S \subseteq [n]\}$ forms an orthonormal basis of $L^2(\{-1, 1\}^n)$, which yields that every $f : \{-1, 1\}^n \to \mathbb{C}$ admits a unique Fourier-Walsh expansion $f = \sum_{S \subseteq [n]} \widehat{f}(S)\, \chi_S$, where $\widehat{f}(S) \in \mathbb{C}$. We say that f is *d-homogeneous* if $\widehat{f}(S) = 0$ whenever $|S| \neq d$, where $|S|$ denotes the cardinality of S.

For any $z \in \mathbb{C}$, the *noise operator T_z* is defined as

$$T_z f = \sum_{S \subseteq [n]} z^{|S|} \widehat{f}(S)\, \chi_S,$$

where the sums is over all subsets $S \subseteq \{1, \ldots, n\}$. When we need to emphasize the dimension of the Boolean cube in question, we will write T_z^n. Let us recall why in computer science T_z is referred to as the noise operator. For $\varepsilon \in [-1, 1]$, given an n-bit string $x \in \{-1, 1\}^n$ define the distribution $y \sim_\varepsilon x$ as follows: independently, each bit of y is equal to the corresponding bit of x with probability $1/2 + \varepsilon/2$, and the bit is flipped with probability $1/2 - \varepsilon/2$. Then a calculation shows that for any $f : \{-1, 1\}^n \to \mathbb{C}$ we have $(T_\varepsilon f)(x) = \mathbb{E}_{y \sim_\varepsilon x}[f(y)]$.

The operator T_z is also known as the *Hermite operator*. When $z = e^{-t}, t \geq 0$, the traditional notation for T_z is $e^{-t\Delta}$ instead of $T_{e^{-t}}$. In the quantum field literature T_z is called the second quantization operator of z.

Since we are considering a probability measure on $\{-1, 1\}^n$, whenever $1 \leq p \leq q \leq \infty$ by Hölder's inequality we have that the identity map $L_q(\{-1, 1\}^n) \to L_p(\{-1, 1\}^n)$ is contractive. The term *hypercontractivity* refers to the fact that for some choices of z, p, q, the operator T_z satisfies the opposite inequality, that is, $T_z : L_p(\{-1, 1\}^n) \to L_q(\{-1, 1\}^n)$ is contractive. A classical strategy for proving this is the "tensor power trick", nicely presented in [43] and which we restate below:

Lemma 5.5.1 *Let* $1 \leq p \leq q \leq \infty$ *and* $z \in \mathbb{C}$ *be fixed. The following are equivalent:*

   (i) *There exists* $C(p, q, z) < \infty$ *such that for all* $n \in \mathbb{N}$, $\|T_z : L_p(\{-1, 1\}^n) \to L_q(\{-1, 1\}^n)\| \leq C(p, q, z)$.
   (ii) *For all* $n \in \mathbb{N}$, $\|T_z : L_p(\{-1, 1\}^n) \to L_q(\{-1, 1\}^n)\| \leq 1$.
   (iii) $\|T_z : L_p(\{-1, 1\}) \to L_q(\{-1, 1\})\| \leq 1$.

Before proceeding to its proof, let us illustrate how Lemma 5.5.1 implies hypercontractivity results in a very straightforward manner.

Corollary 5.5.2 *Let* $1 < p \leq q < \infty$. *Let* $\varepsilon = \sqrt{\frac{p-1}{q-1}}$. *Then for any* $f : \{-1, 1\}^n \to \mathbb{C}$ *we have* $\|T_\varepsilon f\|_q \leq \|f\|_p$. *If* f *is d-homogeneous, then* $\|f\|_q \leq (\frac{q-1}{p-1})^{d/2} \|f\|_p$.

Proof The fact that $\|T_\varepsilon : L_p(\{-1, 1\}) \to L_q(\{-1, 1\})\| \leq 1$ is the famous two-point inequality of Bonami [15], Gross [39] and Beckner [6], so Lemma 5.5.1 immediately gives the first part of the conclusion. The second part follows from the observation that when f is d-homogeneous, $T_\varepsilon f = \varepsilon^d f$. $\qquad\square$

The proof of Lemma 5.5.1 depends on the following multiplicativity result for the $L_p \to L_q$ norms. It is well known that this can be proved using Fubini's theorem and Minkowski's inequality, but we will present a more conceptual proof that has the advantage of being easily adaptable to the operator space setting.

Proposition 5.5.3 *Let* $1 \leq p \leq q < \infty$. *For* $j = 1, 2$, *let* μ_j, ν_j *be measures and* $T_j : L_p(\mu_j) \to L_q(\nu_j)$ *be linear maps. Then*

$$\left\| T_1 \otimes T_2 : L_p(\mu_1 \times \mu_2) \to L_q(\nu_1 \times \nu_2) \right\|$$
$$= \left\| T_1 : L_p(\mu_1) \to L_q(\nu_1) \right\| \left\| T_2 : L_p(\mu_2) \to L_q(\nu_2) \right\|.$$

Proof We will need the following properties of the right Chevet-Saphar Banach space tensor products d_p^B [23, Sec. 12]:

(a) $L_p(\mu_1 \times \mu_2)$ is isometric to $L_p(\mu_1) \widehat{\otimes}_{d_p^B} L_p(\mu_2)$ via the formal identity (and similarly for the L_q's) [23, Sec. 15.10, Cor. 2].
(b) For any Banach spaces X and Y the formal identity map $X \otimes_{d_p^B} Y \to X \otimes_{d_q^B} Y$ is a contraction [23, Prop. 12.5].

Now, by the metric mapping property of d_p^B we have

$$\left\| T_1 \otimes T_2 : L_p(\mu_1) \widehat{\otimes}_{d_p^B} L_p(\mu_2) \to L_q(\nu_1) \widehat{\otimes}_{d_p^B} L_q(\nu_2) \right\|$$
$$\leq \left\| T_1 : L_p(\mu_1) \to L_q(\nu_1) \right\| \left\| T_2 : L_p(\mu_2) \to L_q(\nu_2) \right\|$$

From (b), we get

$$\left\| Id \otimes Id : L_q(\nu_1) \widehat{\otimes}_{d_p^B} L_q(\nu_2) \to L_q(\nu_1) \widehat{\otimes}_{d_q^B} L_q(\nu_2) \right\| \leq 1$$

Composition yields

$$\left\| T_1 \otimes T_2 : L_p(\mu_1) \widehat{\otimes}_{d_p^B} L_p(\mu_2) \to L_q(\nu_1) \widehat{\otimes}_{d_q^B} L_q(\nu_2) \right\|$$
$$\leq \left\| T_1 : L_p(\mu_1) \to L_q(\nu_1) \right\| \left\| T_2 : L_p(\mu_2) \to L_q(\nu_2) \right\|.$$

The reverse inequality is clear (using functions of separated variables), so we must have equality, and this is the desired result thanks to (a). $\qquad\square$

Proof of Lemma 5.5.1 The implications $(ii) \Rightarrow (i)$ and $(ii) \Rightarrow (iii)$ are obvious.

For any $n, m \in \mathbb{N}$, note that the uniform measure on $\{-1, 1\}^{n+m}$ is the product of the uniform measures on $\{-1, 1\}^n$ and $\{-1, 1\}^m$. Therefore, by Proposition 5.5.3 we have

$$\left\| T_z^n \otimes T_z^m : L_p(\{-1, 1\}^{n+m}) \to L_q(\{-1, 1\}^{n+m}) \right\|$$
$$= \left\| T_z^n : L_p(\{-1, 1\}^n) \to L_q(\{-1, 1\}^n) \right\| \left\| T_z^m : L_p(\{-1, 1\}^m) \to L_q(\{-1, 1\}^m) \right\|.$$

For any $S_1 \subseteq [n]$ and $S_2 \subseteq [m]$, observe that

$$(T_z^n \otimes T_z^m)(\chi_{S_1} \otimes \chi_{S_2}) = z^{|S_1|} \chi_{S_1} \otimes z^{|S_2|} \chi_{S_2}$$
$$= z^{|S_1|+|S_2|} \chi_{S_1} \otimes \chi_{S_2} = T_z^{n+m}(\chi_{S_1} \otimes \chi_{S_2}).$$

Since the functions of the form $\chi_{S_1} \otimes \chi_{S_2}$ span all of $L_p(\{-1, 1\}^{n+m})$, we conclude that $T_z^n \otimes T_z^m = T_z^{n+m}$.

Thus, for any $n, k \in \mathbb{N}$ we have

$$\| T_z^n : L_p(\{-1, 1\}^n) \to L_q(\{-1, 1\}^n) \|$$
$$= \| T_z^{nk} : L_p(\{-1, 1\}^{nk}) \to L_q(\{-1, 1\}^{nk}) \|^{1/k},$$

from where $(i) \Rightarrow (ii)$ follows, and also

$$\| T_z^n : L_p(\{-1, 1\}^n) \to L_q(\{-1, 1\}^n) \| = \| T_z^1 : L_p(\{-1, 1\}) \to L_q(\{-1, 1\}) \|^n,$$

from where we get $(iii) \Rightarrow (ii)$. $\square$

In the quantum setting, the classical Boolean cube $\{-1, 1\}^n$ is replaced by the Hilbert space $H_n = (\mathbb{C}^2)^{\otimes n}$ of n-qubits, and the functions $f : \{-1, 1\}^n \to \mathbb{C}$ are replaced by the operators on $(\mathbb{C}^2)^{\otimes n}$, that is, the elements of $M_2(\mathbb{C})^{\otimes n} \cong M_{2^n}(\mathbb{C})$ where $M_k(\mathbb{C})$ denotes the k-by-k matrices with complex entries. The space $L_p(\{-1, 1\}^n)$ is replaced by $L_p(\mathrm{tr}_n)$, the noncommutative L_p space associated to the normalized trace $\mathrm{tr}_n = \frac{1}{2^n} \mathrm{tr}$ on $M_{2^n}(\mathbb{C})$, with norm given by $\|A\|_p = \mathrm{tr}_n(|A|^p)^{\frac{1}{p}}$ where $|A| := (A^*A)^{1/2}$ in the case $1 \le p < \infty$, and with $\| \cdot \|_\infty = \| \cdot \|$ being the usual operator norm. We will additionally consider the standard operator space structure on the spaces $L_p(\mathrm{tr}_n)$ as given by Pisier [64]. We remark that, up to a factor $2^{-\frac{n}{p}}$, this is precisely the space $S_p^{2^n} = S_p[(\mathbb{C}^2)^{\otimes n}]$ from Sects. 1.1.1 and 1.4.

The *Pauli matrices* are the self-adjoint 2×2 unitaries

$$\sigma_0 = \begin{pmatrix} 1 & 0 \\ 0 & 1 \end{pmatrix}, \quad \sigma_1 = \begin{pmatrix} 0 & 1 \\ 1 & 0 \end{pmatrix}, \quad \sigma_2 = \begin{pmatrix} 0 & -i \\ i & 0 \end{pmatrix}, \quad \sigma_3 = \begin{pmatrix} 1 & 0 \\ 0 & -1 \end{pmatrix},$$

which form an orthonormal basis of $L_2(\mathrm{tr}_1)$ with respect to the normalized trace inner product $\langle A, B \rangle = \mathrm{tr}_1(A^*B)$ (note that here we are using the standard physics convention of having the inner products be conjugate linear in the first variable). For an n-qubits system, we use the notation of [11] for the tensor product of Pauli matrices: given a multi-index $s = (s_1, \ldots, s_n) \in \{0, 1, 2, 3\}^n$,

$$\sigma_s := \sigma_{s_1} \otimes \cdots \otimes \sigma_{s_n}. \tag{5.5.1}$$

These form an orthonormal basis of $L_2(\mathrm{tr}_n)$, so every $A \in M_2(\mathbb{C})^{\otimes n}$ has an unique Fourier-Pauli expansion

$$A = \sum_{s \in \{0,1,2,3\}^n} \widehat{A}_s \, \sigma_s,$$

where $\widehat{A}_s = \mathrm{tr}_n(\sigma_s A) \in \mathbb{C}$ are the Pauli coefficients.

Given a multi-index $s = (s_1, \ldots, s_n) \in \{0, 1, 2, 3\}^n$, we define the *support of* s as

$$\mathrm{supp}\, s := \{j \in [n] \mid s_j \neq 0\}$$

and $|\mathrm{supp}\, s|$ denotes its cardinality. We say that $A \in M_2(\mathbb{C})^{\otimes n}$ is *d-homogeneous* if $\widehat{A}_s = 0$ for $|\mathrm{supp}\, s| \neq d$.

Generalizing [51, Def. 8.1] (where only the case of real z was considered) for any $z \in \mathbb{C}$ we define the *noise operator T_z* by

$$T_z(A) = \sum_{s \in \{0,1,2,3\}^n} z^{|\,\mathrm{supp}(s)|} \widehat{A}_s \, \sigma_s, \qquad A \in M_2(\mathbb{C})^{\otimes n}.$$

Once again, when we need to specify the dimension of the quantum Boolean cube involved, we write T_z^n. The abuse of notation is justified by the fact that for matrices which are diagonal in the canonical basis, the classical and quantum versions of T_z agree. In the case $z \in [0, 1]$, the map T_z is known as the *depolarizing channel* in Quantum Information Theory. The maps T_z are also an example of radial multipliers on Fermion algebras in the sense of [4].

Early hypercontractivity results for the operator norm in the quantum Boolean cube setting are due to Montanaro and Osborne [51], who emphasized the fact that one cannot use the tensor product trick because the $L_p \to L_q$ norms are known to not be multiplicative in the noncommutative setting (see also [47] for more general hypercontractivity results). For an operator space specialist, it is immediately tempting to check if the situation is remedied by replacing norms with completely bounded norms. This kind of trade-off is well-established in Quantum Information, for example the $S_1 \to S_1$ norm is not multiplicative but its completely bounded version is [73, Example 3.42 and Theorem 3.49]. Conceptually, the issue is that there is no "nice" Banach space tensor norm that isometrically identifies $S_1^n \otimes S_1^m$ with S_1^{nm}. However, the issue goes away once we consider the operator space structure: now we have the completely isometric identification $S_1^n \otimes_{\mathrm{proj}} S_1^m = S_1^{nm}$.

Indeed, it was shown in the very influential paper [26, Thm. 11] that multiplicativity of the completely bounded $L_p \to L_q$ norms does hold for a special type of maps (the completely positive ones), and this was shown to have various interesting consequences including so-called additivity results in Quantum Information Theory. Throughout the rest of this section we will show how our knowledge of the Chevet-Saphar o.s. tensor norms allows one to recover such a multiplicativity result. See [5, App. 11.2] for an analogous result for more general noncommutative L_p spaces, with a proof from a non-tensorial point of view.

We start with a weaker version valid for general maps, which will then imply the strong version in the completely positive case.

Proposition 5.5.4 *Let* $1 \le p \le q \le \infty$. *For* $j = 1, 2$, *let* H_j, K_j *be Hilbert spaces and* $T_j : S_p(H_j) \to S_q(K_j)$ *be linear maps. Then*

$$\left\| T_1 \otimes T_2 : S_p(H_1 \otimes_2 H_2) \to S_q(K_1 \otimes_2 K_2) \right\|$$

$$\le \left\| T_1 : S_p(H_1) \to S_q(K_1) \right\|_{\mathrm{cb}} \left\| T_2 : S_p(H_2) \to S_q(K_2) \right\|_{\mathrm{cb}}$$

Proof We will need the following properties of the right Chevet-Saphar operator space tensor products d_p:

(a) $S_p(H_1 \otimes_2 H_2)$ is completely isometric to $S_p(H_1) \widehat{\otimes}_{d_p} S_p(H_2)$ via the formal identity, and similarly for the S_q's (Corollary 5.4.5).
(b) For any operator spaces E and F the formal identity map $E \otimes_{d_p} F \to E \otimes_{d_q} F$ is a contraction (Theorem 2.2.7).

It is now evident that the proof of Proposition 5.5.3 can be translated verbatim to this context.   □

To motivate why the completely positive case has advantages which imply the strong multiplicativity result, let us look back at the corresponding situation for positive maps in the classical setting. For example, it is known that for any $1 \le p, q \le \infty$, any measures μ and ν, any normed space X, and a *positive* linear map $T : L_p(\mu) \to L_q(\nu)$ one has

$$\left\| T \otimes Id_X : L_p(\mu; X) \to L_q(\nu; X) \right\| \le \|T\|.$$

(see [23, Thm. 7.3]), but this is not true for general (nonpositive) maps. Let us now prove the corresponding result for a completely positive $T : S_p^n \to S_q^m$ (see [26, Lem. 5] for a special case and the recent paper [38, Thm. 4.1] for an even more general result). Recall that a linear map $T : M_n \to M_m$ is called *positive* if it maps positive semidefinite matrices to positive semidefinite matrices, and *completely positive* if $Id_{M_k} \otimes T : M_{kn} \to M_{km}$ is positive for all $k \in \mathbb{N}$. The completely positive maps have several important characterizations, see e.g. [73, Thm. 2.22]. We will only need the *Kraus representation* [22]: $T : M_n \to M_m$ is completely positive if and only if there exist $A_1, A_2, \ldots, A_r \in M_{mn}$ such that for all $X \in M_n$ we have

$$T(X) = \sum_{j=1}^{r} A_j X A_j^*.$$

Proposition 5.5.5 *Let* $1 \leq p \leq q \leq \infty$. *Let* $T : M_n \to M_m$ *be a completely positive map. Then for any operator space* E,

$$\left\| T \otimes Id_E : S_p^n[E] \to S_q^m[E] \right\| = \left\| T : S_p^n \to S_q^m \right\|.$$

Proof The inequality $\geq$ is obvious, so it will suffice to check the opposite one. Consider a Kraus representation for T as above. Fix $x \in S_p^n[E]$, and write it as $x = a \cdot y \cdot b$ with $a, b \in M_n$ and $y \in M_n(E)$. It follows that

$$(T \otimes Id_E)x = \sum_{j=1}^{r} A_j a \cdot y \cdot b A_j^* = \begin{bmatrix} A_1 a & A_2 a & \cdots & A_r a \end{bmatrix} \begin{bmatrix} y & 0 & 0 & \cdots & 0 \\ 0 & y & 0 & \cdots & 0 \\ \vdots & & & & \vdots \\ 0 & 0 & 0 & \cdots & y \end{bmatrix} \begin{bmatrix} bA_1^* \\ bA_2^* \\ \vdots \\ bA_r^* \end{bmatrix}.$$

Thus, denoting $\widetilde{A} = \begin{bmatrix} A_1 a & A_2 a & \cdots & A_r a \end{bmatrix}$ and $\widetilde{B} = \begin{bmatrix} bA_1^* \\ bA_2^* \\ \vdots \\ bA_r^* \end{bmatrix}$ we have by

Theorem 1.4.8

$$\begin{aligned}
\left\| (T \otimes Id_E)x \right\|_{S_q^m[E]} &\leq \left\| \widetilde{A} \right\|_{S_{2q}} \left\| I_r \otimes y \right\|_{M_{nr}(E)} \left\| \widetilde{B} \right\|_{S_{2q}} \\
&= \left\| \widetilde{A}\widetilde{A}^* \right\|_{S_q}^{1/2} \left\| y \right\|_{M_n(E)} \left\| \widetilde{B}^*\widetilde{B} \right\|_{S_q}^{1/2} \\
&= \left\| \sum_{j=1}^{r} A_j aa^* A_j^* \right\|_{S_q}^{1/2} \left\| y \right\|_{M_n(E)} \left\| \sum_{j=1}^{r} A_j b^* b A_j^* \right\|_{S_q}^{1/2} \\
&= \left\| T(aa^*) \right\|_{S_q}^{1/2} \left\| y \right\|_{M_n(E)} \left\| T(b^*b) \right\|_{S_q}^{1/2} \\
&\leq \left\| T : S_p^n \to S_q^m \right\| \left\| aa^* \right\|_{S_p}^{1/2} \left\| y \right\|_{M_n(E)} \left\| b^*b \right\|_{S_p}^{1/2} \\
&= \left\| T : S_p^n \to S_q^m \right\| \left\| a \right\|_{S_{2p}} \left\| y \right\|_{M_n(E)} \left\| b \right\|_{S_{2p}}.
\end{aligned}$$

Taking the infimum over all representations $x = a \cdot y \cdot b$, by Theorem 1.4.8 we get the desired conclusion.   □

We now bootstrap the previous result to show that for the completely positive maps $S_p \to S_q$ with $p \leq q$, the norm and the completely bounded norm coincide, which essentially explains why in this case the weak multiplicativity result of Proposition 5.5.4 will imply the strong one.

Proposition 5.5.6 *Let* $1 \leq p \leq q \leq \infty$.

(a) Let $T : M_n \to M_m$ *be a completely positive map. Then*

$$\left\| T : S_p^n \to S_q^m \right\|_{\mathrm{cb}} = \left\| T : S_p^n \to S_q^m \right\|.$$

(b) More generally, if H_1, H_2 *are Hilbert spaces and* $T : S_p(H_1) \to S_q(H_2)$ *is completely positive, then*

$$\left\| T : S_p(H_1) \to S_q(H_2) \right\|_{\mathrm{cb}} = \left\| T : S_p(H_1) \to S_q(H_2) \right\|.$$

Proof

(a) The inequality $\geq$ is obvious, so it will suffice to check the opposite one. Denote $C = \left\| T : S_p^n \to S_q^m \right\|$. By Theorem 1.4.3, it suffices to show that for every $k \in \mathbb{N}$ we have

$$\left\| Id_k \otimes T : S_q^k[S_p^n] \to S_q^k[S_q^m] \right\| \leq C. \tag{5.5.2}$$

First we will prove that for any $a, b \in M_k$,

$$\left\| M(a,b) \otimes T : S_p^k[S_p^n] \to S_q^k[S_q^m] \right\| \leq C \left\| a \right\|_{S_{2r}^k} \left\| b \right\|_{S_{2r}^k}. \tag{5.5.3}$$

Indeed, write it as the composition

$$S_p^k[S_p^n] \to S_p^n[S_p^k] \xrightarrow{T \otimes Id_k} S_q^m[S_p^k] \xrightarrow{Id_m \otimes M(a,b)} S_q^m[S_q^k] \to S_q^k[S_q^m],$$

where the first and last arrows are the flip maps, which are isometries by the Fubini theorem for Schatten spaces. Since T is completely positive, by Proposition 5.5.5 the second arrow has norm C. The third arrow has norm at most $\left\| a \right\|_{S_{2r}^m} \left\| b \right\|_{S_{2r}^m}$ by Theorems 1.4.3 and 1.4.7.

Now, let $x \in S_q^k[S_p^n]$. Write $x = a \cdot y \cdot b = (M(a,b) \otimes Id_n)y$ for some $a, b \in M_k, y \in M_k(S_p^n)$. Then, by (5.5.3),

$$\left\| (Id_k \otimes T)x \right\|_{S_q^k[S_q^m]} = \left\| (M(a,b) \otimes T)y \right\|_{S_q^k[S_q^m]} \leq C \left\| a \right\|_{S_{2r}^m} \left\| b \right\|_{S_{2r}^m} \left\| y \right\|_{S_p^k[S_p^n]}.$$

Taking the infimum over all such representations of x, Theorem 1.4.8 implies

$$\left\| (Id_k \otimes T)x \right\|_{S_q^k[S_q^m]} \leq C \left\| x \right\|_{S_q^k[S_p^n]},$$

yielding the desired inequality (5.5.2).

(*b*) For $j = 1, 2$ let $K_j \subset H_j$ be finite-dimensional subspaces, and let $V_j : K_j \to H_j$ be the inclusion maps. Define

$$\tilde{T} : S_p(K_1) \to S_q(K_2), \qquad \tilde{T}(X) = V_2^*\big[T(V_1 X V_1^*)\big] V_2,$$

which should be understood as restricting T to the subspace $S_p(K_1) \subseteq S_p(H_1)$, and then compressing its values from $S_q(H_1)$ to $S_q(K_2)$. Note that $\tilde{T}$ is completely positive since it is a composition of completely positive maps. Therefore, by part (*a*) we have

$$\big\| \tilde{T} : S_p(K_1) \to S_q(K_2) \big\|_{\mathrm{cb}} = \big\| \tilde{T} : S_p(K_1) \to S_q(K_2) \big\| .$$

Since the set $\bigcup_{K_1 \subset H_1} S_p(K_1)$ (where K_1 ranges over all possible finite-dimensional subspaces of H_1) is dense in $S_p(H_1)$, and a similar statement holds for $\bigcup_{K_2 \subset H_2} S_q(K_2)$, by an approximation argument we obtain the desired conclusion.

$\square$

Corollary 5.5.7 *Let $1 \le p \le q \le \infty$. For $j = 1, 2$, let H_j, K_j be Hilbert spaces and $T_j : S_p(H_j) \to S_q(K_j)$ be completely positive linear maps. Then*

$$\big\| T_1 \otimes T_2 : S_p(H_1 \otimes_2 H_2) \to S_q(K_1 \otimes_2 K_2) \big\|_{\mathrm{cb}}$$

$$\le \big\| T_1 : S_p(H_1) \to S_q(K_1) \big\|_{\mathrm{cb}} \big\| T_2 : S_p(H_2) \to S_q(K_2) \big\|_{\mathrm{cb}}$$

Proof From Proposition 5.5.4 we already know $\| T_1 \otimes T_2 \| \le \| T_1 \|_{\mathrm{cb}} \| T_2 \|_{\mathrm{cb}}$. Since T_1 and T_2 are completely positive so is $T_1 \otimes T_2$, and hence applying Proposition 5.5.6 yields the desired result.

$\square$

Going back to the quantum Boolean cube, it is also clear that now one gets the completely bounded version of the "tensor product trick" in Lemma 5.5.8 below, remembering that $L_p(\mathrm{tr}_n)$ is completely isometrically identified with $2^{-\frac{n}{p}} S_p[(\mathbb{C}^2)^{\otimes n}]$ via the formal identity. The key is that evaluating on operators of the form σ_s (see Eq. (5.5.1)) shows that one can identify

$$T_z^{n+m} : M_2(\mathbb{C})^{\otimes (n+m)} \to M_2(\mathbb{C})^{\otimes (n+m)}$$

with the tensor product

$$T_z^n \otimes T_z^m : M_2(\mathbb{C})^{\otimes n} \otimes M_2(\mathbb{C})^{\otimes m} \to M_2(\mathbb{C})^{\otimes n} \otimes M_2(\mathbb{C})^{\otimes m}.$$

Lemma 5.5.8 *Let $1 \le p \le q \le \infty$ and $z \in [0, 1]$ be fixed. The following are equivalent:*

(i) *There exists $C(p, q, z) < \infty$ such that for all $n \in \mathbb{N}$, $\| T_z \cdot L_p(\mathfrak{u}_n) \to L_q(\mathrm{tr}_n) \|_{\mathrm{cb}} \le C(p, q, z)$.*

(ii) For all $n \in \mathbb{N}$, $\|T_z : L_p(\mathrm{tr}_n) \to L_q(\mathrm{tr}_n)\|_{\mathrm{cb}} \leq 1$.
(iii) $\|T_z : L_p(\mathrm{tr}_1) \to L_q(\mathrm{tr}_1)\|_{\mathrm{cb}} \leq 1$.

Proof Evaluating on the Pauli matrices shows that for any $A \in M_2(\mathbb{C})$ and any $z \in \mathbb{C}$ we have

$$T_z^1(A) = zA + \frac{1-z}{2}\,\mathrm{tr}(A)I_2$$

Having $z \in [0, 1]$ thus implies that T_z^1 is completely positive, since in this case $T_z^1 : M_2(\mathbb{C}) \to M_2(\mathbb{C})$ is a linear combination with nonnegative coefficients of two maps which are completely positive: the identity and the map $A \mapsto \mathrm{tr}(A)I_2$.

The same proof as in Lemma 5.5.1 now applies, using Corollary 5.5.7 instead of Proposition 5.5.3. $\qquad\square$

Hypercontractivity in the completely bounded setting has certainly already been considered, see for example [8] where it is shown that (for certain semigroups) it is equivalent to a completely bounded version of the log-Sobolev inequality.

While in principle it may seem too much to ask to hope for a version of Corollary 5.5.2 in the completely bounded case with exactly the same parameters, it turns out that this is in fact the case.

Corollary 5.5.9 *Let $1 < p \leq q < \infty$. Let $\varepsilon = \sqrt{\frac{p-1}{q-1}}$. Then for any $n \in \mathbb{N}$, $\|T_\varepsilon^n : L_p(\mathrm{tr}_n) \to L_q(\mathrm{tr}_n)\|_{\mathrm{cb}} \leq 1$. If $(A_{ij})_{i,j=1}^m$ is a matrix of d-homogeneous operators $A_{ij} \in M_2(\mathbb{C})^{\otimes n}$, then*

$$\left\|(A_{ij})\right\|_{M_m(L_q(\mathrm{tr}_n))} \leq \left(\frac{q-1}{p-1}\right)^{d/2} \left\|(A_{ij})\right\|_{M_m(L_p(\mathrm{tr}_n))}.$$

Proof By Lemma 5.5.8, it suffices to prove the result for $n = 1$. Since T_ε^1 is completely positive (as explained in the proof of Lemma 5.5.8), by Proposition 5.5.6 it suffices to show that $\left\|T_\varepsilon : L_p(\mathrm{tr}_1) \to L_q(\mathrm{tr}_1)\right\| \leq 1$. This is precisely [51, Prop. 8.5].

The second part of the conclusion is obtained analogously to what was done in Corollary 5.5.2. $\qquad\square$

Beyond the application above, let us now show that the monotonicity of the Chevet-Saphar o.s. tensor norms implies a multiplicativity result which is more general than Proposition 5.5.4. See [23, Prop. 15.12] for the classical analogue.

Theorem 5.5.10 *Let $1 \leq p \leq q \leq \infty$ and $T_1 : E \to S_p(H)$, $T_2 : S_q(K) \to F$ be completely bounded, where H and K are Hilbert spaces. Then*

$$\left\|T_1 \otimes T_2 : E \otimes_{\Delta_p^t} S_p(K) \to S_q(H) \otimes_{\Delta_q} F\right\| \leq \|T_1\|_{\mathrm{cb}} \|T_2\|_{\mathrm{cb}},$$

Proof The map under consideration can be written as the composition

$$E \otimes_{\Delta_p^t} S_p(K) \xrightarrow{(a)} E \otimes_{g_p} S_p(K) \xrightarrow{T_1 \otimes T_2} S_q(H) \otimes_{g_p} F$$

$$\xrightarrow{(b)} S_q(H) \otimes_{g_{p'}^*} F \xrightarrow{(c)} S_q(H) \otimes_{g_{q'}^*} F \xrightarrow{(d)} S_q(H) \otimes_{\Delta_q} F$$

where the arrows (a), (b), (c), (d) are just the formal identities. The arrow $T_1 \otimes T_2$ has cb-norm less than $\|T_1\|_{cb} \|T_2\|_{cb}$ by the uniformity of the o.s. tensor norm g_p. The arrows (a), (b) and (d) are completely contractive by the Chevet-Persson-Saphar inequalities (Theorem 5.4.4), whereas (c) is contractive by Theorem 2.2.7.

$\square$

In particular, replacing E (resp. F) by $S_p[E]$ (resp. $S_q[F]$) yields the following:

Corollary 5.5.11 *Let $1 \leq p \leq q \leq \infty$ and $T_1 : S_p[K_1; E] \to S_q(H_1)$ and $T_2 : S_p(K_2) \to S_q[H_2; F]$ be completely bounded. Then*

$$\left\| T_1 \otimes T_2 : S_p[K_1 \otimes_2 K_2; E] \to S_q[H_1 \otimes_2 H_2; F] \right\|$$

$$\leq \left\| T_1 : S_p[K_1; E] \to S_q(H_1) \right\|_{cb} \left\| T_2 : S_p(K_2) \to S_q[H_2; F] \right\|_{cb}.$$

We finish the section by mentioning a couple of other generalizations of multiplicativity results in Quantum Information which follow easily from the o.s. tensor norm point of view, and which are valid even for maps that are not necessarily completely positive. The proofs are quite similar to those of [38], with the fundamental difference that we are using the general theory of o.s. tensor norms to justify the steps. First, a generalization of [38, Cor. 4.8], which in turn yields a special case of the generalized Entropy Accumulation Theorem of [50] (see [38] for the details).

Lemma 5.5.12 *Let $1 \leq p < \infty$. For any Hilbert spaces H_1, H_2, K, operator spaces E and F, and linear maps $T_1 : S_1(H_1) \to S_1(H_2)$ and $T_2 : E \to F$,*

$$\left\| T_1 \otimes Id_{S_p(K)} \otimes T_2 : S_1[H_1; S_p[K; E]] \to S_1[H_2; S_p[K; F]] \right\|_{cb}$$

$$\leq \|T_1\|_{cb} \|T_2\|_{cb}.$$

Proof Since the vector valued S_1 spaces are defined by taking the proj tensor product, using the uniformity of proj and Theorem 1.4.3 we have

$$\left\| T_1 \otimes Id_{S_p(K)} \otimes T_2 : S_1(H_1) \otimes_{\mathrm{proj}} S_p[K; E] \to S_1(H_2) \otimes_{\mathrm{proj}} S_p[K; F] \right\|_{cb}$$

$$\leq \|T_1\|_{cb} \left\| Id_{S_p(K)} \otimes T_2 : S_p[K; E] \to S_p[K; F] \right\|_{cb} \leq \|T_1\|_{cb} \|T_2\|_{cb}.$$

$\square$

Next, a generalization of a particular case of [38, Thm. 4.10]. While generalizing only a particular case sounds limited, it is worth pointing out that the particular case which we generalize is precisely the one used to prove the additivity of output α-Rényi conditional entropy [38, Eqn. (2)].

Lemma 5.5.13 *Let $1 \leq p \leq \infty$. For $j = 1, 2$ suppose H_j, K_j, G_j are Hilbert spaces and $T_j : S_1(H_j) \to S_1[K_j, S_p(G_j)]$ are linear maps. Then*

$$\left\| T_1 \otimes T_2 : S_1(H_1 \otimes_2 H_2) \to S_1[K_1 \otimes_2 K_2; S_p(G_1 \otimes_2 G_2)] \right\|_{\mathrm{cb}} \leq \|T_1\|_{\mathrm{cb}} \|T_2\|_{\mathrm{cb}}.$$

Proof Once again, the two key ingredients are first that vector valued S_1 spaces are given by a proj tensor product and, second identification from Theorem 1.4.5. Consider the composition

$$S_1(H_1) \otimes_{\mathrm{proj}} S_1(H_2)$$

$$\xrightarrow{T_1 \otimes T_2} \left(S_1(K_1) \otimes_{\mathrm{proj}} S_p(G_1) \right) \otimes_{\mathrm{proj}} \left(S_1(K_2) \otimes_{\mathrm{proj}} S_p(G_2) \right)$$

$$\xrightarrow{\text{shuffle}} \left(S_1(K_1) \otimes_{\mathrm{proj}} S_1(K_2) \right) \otimes_{\mathrm{proj}} \left(S_p(G_1) \otimes_{\mathrm{proj}} S_p(G_2) \right)$$

$$\xrightarrow{(a)} \left(S_1(K_1) \otimes_{\mathrm{proj}} S_1(K_2) \right) \otimes_{\mathrm{proj}} \left(S_p(G_1) \otimes_{d_p} S_p(G_2) \right)$$

$$\xrightarrow{(b)} S_1(K_1 \otimes_2 K_2) \otimes_{\mathrm{proj}} S_p(G_1 \otimes_2 G_2),$$

where the arrows (a) and (b) are the formal identities. The arrow $T_1 \otimes T_2$ has cb-norm $\|T_1\|_{\mathrm{cb}} \|T_2\|_{\mathrm{cb}}$ by the uniformity of proj. The shuffle is a complete isometry by the commutativity and associativity of proj. The arrow (a) is completely contractive because of the uniformity of proj and the fact that $d_p \leq$ proj. The arrow (b) is completely contractive by the Fubini theorem for d_p (Corollary 5.4.5) and the uniformity of proj once more. $\square$

Chapter 6
The Completely Bounded Approximation Property

Recall that given $C \geq 1$, a normed operator space is said to have the *C-completely bounded approximation property* (*C*-CBAP for short) if there exists a net of finite-rank mappings $\phi_i : E \to E$ such that $\|\phi_i\|_{\mathrm{cb}} \leq C$ for all i and for every $x \in E$, $\|\phi_i(x) - x\| \to 0$. One of the goals in this chapter is to study the CBAP via various conditions involving tensor products, in the spirit of the characterizations for the operator space approximation property appearing in [34, Sec. 11.2]. We also deal with a weaker version, called W*CBAP, which carries over many of the interesting equivalences that appear in the classical theory of tensor products [23, Sec. 16].

For a normed operator space E, recall we use the notation $K_\infty(E) = \mathcal{K} \widehat{\otimes}_{\min} E$ to emphasize that we think of this space as consisting of infinite E-valued matrices; see [34, Sec. 10.1] for the specific details of how to interpret $K_\infty(E)$ as a completion of the union of the spaces $M_n(E)$ of E-valued matrices. Note that this is exactly the same as the vector valued Schatten space $S_\infty[E]$, but in this chapter we will use the notation $K_\infty(E)$ to facilitate the comparison with the related results from [31]. Given normed operator spaces E and F, recall that a net (φ^α) in $\mathrm{CB}(E, F)$ converges to φ in the *stable point-norm topology* if for every $x \in K_\infty(E)$ we have that $\varphi_\infty^\alpha(x) \to \varphi_\infty(x)$ in the norm topology of $K_\infty(F)$ [34, Sec. 11.2].

6.1 Tensor Product Characterizations

Our first characterization of CBAP (Proposition 6.1.2 below) corresponds to [34, Lem. 11.2.1], but instead of the stable point-norm topology we simply use the *point-norm topology* (which we denote by τ). The reason is clarified in the following remark.

Remark 6.1.1 Observe that for a bounded net in $\mathrm{CB}(E, F)$, convergence in the stable point-norm topology coincides with convergence in the point-norm topology. Indeed, suppose $\varphi^\alpha, \varphi \subset \mathrm{CB}(E, F)$ and $x \in K_\infty(E)$. If $P^m : K_\infty(E) \to K_\infty(E)$

J. A. Chávez-Domínguez et al., *Operator Space Tensor Norms*, Lecture Notes in Mathematics 2379, https://doi.org/10.1007/978-3-031-96209-7_6

denotes the truncation operator,

$$\left\| \varphi_\infty^\alpha(x) - \varphi_\infty(x) \right\|_{K_\infty(F)} \leq \left\| \varphi^\alpha \right\|_{\mathrm{cb}} \left\| x - P^m(x) \right\|_{K_\infty(E)}$$

$$+ \left\| \varphi_m^\alpha(P^m(x)) - \varphi_m(P^m(x)) \right\|_{M_m(F)} + \left\| \varphi \right\|_{\mathrm{cb}} \left\| P^m(x) - x \right\|_{K_\infty(E)},$$

which shows that when φ^α is bounded, convergence in the point-norm topology implies convergence in the stable point-norm topology (the other implication is trivial and holds in general).

Proposition 6.1.2 *Suppose that E is a normed operator space and $C \geq 1$. The following are equivalent:*

(a) E has the C-CBAP.
(b) $C \cdot B_{E' \otimes_{\min} E}$ is τ-dense in $B_{\mathrm{CB}(E,E)}$.
(c) For any normed operator space F, $C \cdot B_{E' \otimes_{\min} F}$ is τ-dense in $B_{\mathrm{CB}(E,F)}$.
(d) For any normed operator space F, $C \cdot B_{F' \otimes_{\min} E}$ is τ-dense in $B_{\mathrm{CB}(F,E)}$.

Proof $(a) \Rightarrow (b)$: Suppose that E has C-CBAP, so that there exists a net of finite-rank mappings $\phi_i : E \to E$ such that $\|\phi_i\|_{\mathrm{cb}} \leq C$ for all i and for every $x \in E$, $\|x - \phi_i x\| \to 0$. This precisely means that $\phi_i \xrightarrow{\tau} id_E$, and note that each ϕ_i can be identified with an element of $C \cdot B_{E' \otimes_{\min} E}$. For any other $\varphi \in B_{\mathrm{CB}(E,E)}$, the net $(\phi_i \circ \varphi)_i$ does the job. Reversing the argument gives $(b) \Rightarrow (a)$.

Now, if (ϕ_i) is a net in $C \cdot B_{E' \otimes_{\min} E}$ converging to id_E in the τ topology, then for any $\psi \in B_{\mathrm{CB}(E,F)}$ the net $\psi \circ \phi_i \in C \cdot B_{E' \otimes_{\min} F}$ converges to ψ in the τ topology. Similarly, for $\psi \in B_{\mathrm{CB}(F,E)}$ the net $\phi_i \circ \psi \in C \cdot B_{F' \otimes_{\min} E}$ converges to ψ in the τ topology. This proves $(b) \Rightarrow (c)$ and $(b) \Rightarrow (d)$. The implications $(c) \Rightarrow (b)$ and $(d) \Rightarrow (b)$ are clear by specializing to $F = E$. $\qquad\square$

It should be noted that although completeness plays an important role in [34, Sec. 11.2], we do not assume it. This was to be expected, as the same is true of the Banach space case: note that in [23] the section on BAP deals with normed spaces, whereas the one on AP deals with Banach spaces. In this regard we point out the following easy lemma, which shows that when it comes to the CBAP it makes no difference to work with a space or with its completion. As usual, $\overline{E}$ denotes the completion of E. Moreover, given $\varphi \in \mathrm{CB}(E, F)$ we denote by $\widetilde{\varphi}$ its unique extension to a map in $\mathrm{CB}(\overline{E}, \overline{F})$.

Lemma 6.1.3 *E has C-CBAP if and only if $\overline{E}$ has C-CBAP.*

Proof

$(\Rightarrow)$ Let (φ^α) be a net of finite-rank maps in $\mathrm{CB}(E, E)$ of cb-norm at most C converging to the identity of E in the point-norm topology. Let $\widetilde{x} \in \overline{E}$. By a triangle inequality argument, for $x \in E$

$$\left\| \widetilde{\varphi}^\alpha(\widetilde{x}) - \widetilde{x} \right\| \leq \left\| \widetilde{\varphi}^\alpha \right\| \|\widetilde{x} - x\| + \left\| \varphi^\alpha(x) - x \right\| + \|x - \widetilde{x}\|,$$

which shows $(\widetilde{\varphi}^{\alpha})$ is a net of finite-rank maps of cb-norm at most C converging to the identity of E in the point-norm topology.

($\Leftarrow$)   Let (ψ^{α}) be a net of finite-rank maps in $\mathrm{CB}(\overline{E}, \overline{E})$ of cb-norm at most C converging to the identity of $\overline{E}$ in the point-norm topology. For each α let E_{α} be the range of ψ^{α}, and for each $\delta > 0$ use perturbation (as in [65, Lem. 2.13.2]) to find a linear map $R_{\alpha,\delta} : E_{\alpha} \to E$ with $\left\| R_{\alpha,\delta} - Id_{E_{\alpha}} \right\|_{\mathrm{cb}} < \delta$. Let $\varphi^{\alpha,\delta}$ be the restriction to E of $\frac{1}{1+\delta} R_{\alpha,\delta} \circ \psi^{\alpha}$. Then for any $x \in E$,

$$\left\| \varphi^{\alpha,\delta}(x) - x \right\|$$

$$\leq \frac{1}{1+\delta} \left\| R_{\alpha,\delta} \psi^{\alpha}(x) - \psi^{\alpha}(x) \right\| + \frac{\delta}{1+\delta} \left\| \psi^{\alpha}(x) \right\| + \left\| \psi^{\alpha}(x) - x \right\|,$$

so $(\varphi^{\alpha,\delta})$ is a net of finite-rank maps in $\mathrm{CB}(E, E)$ of cb-norm at most C converging to the identity of E in the point-norm topology.

$\square$

For normed operator spaces E and F, recall that a net (φ^{α}) in $\mathrm{CB}(E, F)$ converges to φ in the *stable point-weak topology* (denoted τ_w) if for every $x \in K_{\infty}(E)$ we have that $\varphi^{\alpha}_{\infty}(x) \to \varphi_{\infty}(x)$ in the weak topology of $K_{\infty}(F)$ [34, Sec. 11.2].

Let us remark that although [34, Sec. 11.2] works with Banach operator spaces, the proof of [34, Prop. 11.2.2] applies more generally to normed operator spaces. Therefore, the proof of [34, Cor. 11.2.3] together with Proposition 6.1.2 yield the following result.

Corollary 6.1.4 *Suppose that E is a normed operator space and $C \geq 1$. The following are equivalent:*

(a) E has the C-CBAP.
(b) $C \cdot B_{E' \otimes_{\min} E}$ is τ_w-dense in $B_{\mathrm{CB}(E,E)}$.
(c) For any operator space F, $C \cdot B_{E' \otimes_{\min} F}$ is τ_w-dense in $B_{\mathrm{CB}(E,F)}$.
(d) For any operator space F, $C \cdot B_{F' \otimes_{\min} E}$ is τ_w-dense in $B_{\mathrm{CB}(F,E)}$.

Consider now the natural embedding

$$\mathrm{CB}(E, F) \subseteq \mathrm{CB}(E, F'') = (E \otimes_{\mathrm{proj}} F')'.$$

In [34, Lem. 11.2.4] it is proved that when E and F are complete, the stable point-weak topology on $\mathrm{CB}(E, F)$ is just the relative weak* topology determined by $E \otimes_{\mathrm{proj}} F'$, and thus each stable point-weakly continuous functional on $\mathrm{CB}(E, F)$ is determined by an element of $E \widehat{\otimes}_{\mathrm{proj}} F'$. Let us once again remark that the proof of [34, Lem. 11.2.4] applies more generally to normed operator spaces.

The bounded approximation property, in the context of normed spaces, has many important equivalences in the tensor product realm. Indeed, a Banach space E has the C-approximation property if and only if $E \otimes_{\pi} F \to (E' \otimes_{\varepsilon} F')'$ is a

C-isomorphism onto its image. Moreover, these properties are also equivalent to the following inequality: $\pi \leq C\overleftarrow{\pi}$ on $E \otimes F$ (for all normed spaces F). At first glance the CBAP seems to be a nice extension of the bounded approximation property to operator spaces, and the literature shows a number of examples where the analogy works very well. However, the lack of local reflexivity does not directly allow the equivalence between CBAP and the natural operator space versions of the mentioned results related to tensor products. Nevertheless, we will see that a weak version of CBAP introduced in [31] will allow us to translate those approximation properties relative to tensor products of normed spaces into this context.

6.2   Weak Approximations

The following is the obvious adaptation of the notion of W*MAP defined in [31].

Definition 6.2.1 An operator space E has the C-W*CBAP if there exists a net of finite-rank maps $\phi_i : E \to E''$ such that $\|\phi_i\|_{\mathrm{cb}} \leq C$ for all i and for every $x \in E$, $\phi_i(x) \to \kappa_E(x)$ in the weak* topology.

Note that clearly the C-CBAP implies the C-W*CBAP, as convergence in norm implies convergence in the weak* topology. The same argument as in [31] shows that if E has the C-W*CBAP, then so does $M_n(E)$.

Remark 6.2.2 Similarly as in [31] it is easy to prove (taking adjoints) that E has the C-CBAP if and only if there is a net of weak* continuous finite-rank maps $\psi_i : E' \to E'$ such that $\|\psi_i\|_{\mathrm{cb}} \leq C$ for all i and $\psi_i(x') \to x'$ in the weak* topology for every $x' \in E'$. Instead, E has the C-W*CBAP if and only if the same condition as above holds with the exception that the weak* continuity of the mappings ψ_i is not assumed/required.

We now prove a variant of Corollary 6.1.4.

Proposition 6.2.3 *Suppose that E is a normed operator space and $C \geq 1$. The following are equivalent:*

*(a) E has the C-W*CBAP.*
(b) $C \cdot B_{E' \otimes_{\min} E''}$ is point-weak-dense in $B_{\mathrm{CB}(E,E'')}$.*
(c) For any operator space F, $C \cdot B_{E' \otimes_{\min} F'}$ is point-weak-dense in $B_{\mathrm{CB}(E,F')}$.*

Proof The implication $(a) \Rightarrow (c)$ is implicit in the proof of [31, Thm. 2.2]. Suppose that $T : E \to F'$ is a complete contraction. Consider the commutative diagram

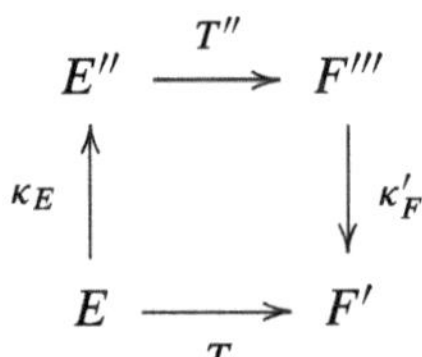

Note that both T'' and κ'_F are weak* continuous. Therefore, if $\phi_i : E \to E''$ is the net of maps in $C \cdot B_{E' \otimes_{\min} E''}$ converging to κ_E in the point-weak* topology given by the definition of C-W*CBAP, it is clear that $\kappa'_F T'' \phi_i : E \to F'$ converges to T in the point-weak* topology.

$(c) \Rightarrow (b)$ is trivial.

Finally, (b) implies that the canonical inclusion $\kappa_E : E \to E''$ can be approximated in the point-weak* topology by a net of finite-rank maps whose cb-norms are at most C, which is exactly the definition of C-W*CBAP, yielding (a).

$\square$

The following result is, in a sense, a quantitative version of [31, Thm. 2.2].

Theorem 6.2.4 *Let E be an operator space and $C \geq 1$. The following are equivalent:*

(a) *For any operator space F, the natural map $E \otimes_{\mathrm{proj}} F \to (E' \otimes_{\min} F')'$ is a complete C-isomorphism onto its image.*

(b) *For any operator space F, the natural map $E \otimes_{\mathrm{proj}} F \to (E' \otimes_{\min} F')'$ is a C-isomorphism onto its image.*

(c) *E has the C-W*CBAP.*

(d) *For any complete contraction $T : E \to E''$ there exists a net of finite-rank maps $\phi_i : E \to E''$ such that $\|\phi_i\|_{\mathrm{cb}} \leq C$ for all i and for every $x \in E$, $\phi_i(x) \to T(x)$ in the weak* topology.*

(e) *For any operator space F and any complete contraction $T : E \to F'$, there is a net of finite-rank maps $\psi_i : E \to F'$ such that $\|\psi_i\|_{\mathrm{cb}} \leq C$ for all i and for every $x \in E$, $\psi_i(x) \to T(x)$ in the weak* topology.*

(f) *The natural map $E \otimes_{\mathrm{proj}} E' \to (E' \otimes_{\min} E'')'$ is a complete C-isomorphism onto its image.*

(g) *The natural map $E \otimes_{\mathrm{proj}} E' \to (E' \otimes_{\min} E'')'$ is a C-isomorphism onto its image.*

Proof $(a) \Rightarrow (b)$ is obvious. Let us assume (b), and let F be an operator space. In order to show that the natural map $E \otimes_{\mathrm{proj}} F \to (E' \otimes_{\min} F')'$ is a complete C-isomorphism onto its image, it suffices to show that its adjoint $(E' \otimes_{\min} F')'' \to (E \otimes_{\mathrm{proj}} F)' = \mathrm{CB}(E, F')$ is a complete C-quotient. In turn, this will follow from proving that for any n the amplification $M_n((E' \otimes_{\min} F')'') \to M_n(\mathrm{CB}(E, F'))$ is a C-quotient. Considering the standard identifications

$$M_n((E' \otimes_{\min} F')'') = (M_n(E' \otimes_{\min} F'))'' = (E' \otimes_{\min} M_n(F'))'' = (E' \otimes_{\min} S_1^n[F]')''$$

and $M_n(\mathrm{CB}(E, F')) = \mathrm{CB}(E, M_n(F')) = \mathrm{CB}(E, S_1^n[F]')$, we need to show that the natural map $(E' \otimes_{\min} (S_1^n[F])')'' \to \mathrm{CB}(E, S_1^n[F]')$ is a C-quotient. This follows from (b) with $S_1^n[F]$ in place of F and taking adjoints. Thus, $(a) \Leftrightarrow (b)$.

Now, $(c) \Leftrightarrow (d) \Leftrightarrow (e)$ is just a restatement of Proposition 6.2.3.

Let us now prove that $(b) \Leftrightarrow (e)$. First observe that the natural map $E \otimes_{\mathrm{proj}} F \to (E' \otimes_{\min} F')'$ always has cb norm one. Therefore, (b) is equivalent to having

$$(E \otimes F) \cap B_{(E' \otimes_{\min} F')'} \subset C \cdot B_{E \otimes_{\mathrm{proj}} F}. \tag{6.2.1}$$

Let us recall the following simple fact stated in [23, p. 191]: if $\langle V, W \rangle$ is a separating dual system, $A, B \subset V$, and $C > 0$, then $A \subset C B^{\circ\circ}$ if and only if $B^{\circ} \subset C A^{\circ}$. For the choices $V = \mathrm{CB}(E, F')$, $W = E \otimes_{\mathrm{proj}} F$, $A = B_{\mathrm{CB}(E,F')}$ and $B = B_{E' \otimes_{\min} F'}$, note that $A^{\circ} = B_{E \otimes_{\mathrm{proj}} F}$ and $B^{\circ} = (E \otimes F) \cap B_{(E' \otimes_{\min} F')'}$, so by the aforementioned fact and the bipolar theorem we get that (6.2.1) is equivalent to

$$B_{\mathrm{CB}(E,F')} \subset C \cdot \overline{B_{E' \otimes_{\min} F'}}^{\,\sigma(\mathrm{CB}(E,F'),\, E \otimes_{\mathrm{proj}} F)}.$$

Since the $\sigma(\mathrm{CB}(E, F'), E \otimes_{\mathrm{proj}} F)$-topology on $\mathrm{CB}(E, F')$ is simply the point-weak* topology, this is clearly equivalent to the statement of (e).

Finally, it is evident that $(a) \Rightarrow (f) \Rightarrow (g)$ and arguing as in $(b) \Leftrightarrow (e)$ we easily obtain that $(g) \Leftrightarrow (d)$. $\qquad\square$

Remark 6.2.5 In view of our preceding results, it seems plausible to extend the usual concept of the bounded approximation property with respect to a tensor norm (see e.g. [23, 21.7]) to our operator space setting. Namely, a natural definition to consider should be the following: given an o.s. tensor norm α, an operator space E has the α-W*CBAP (with constant $C > 0$) if the natural map

$$F \otimes_{\alpha} E \to (F' \otimes_{\alpha'} E')'$$

is a complete C-isomorphism onto its image for every operator space F. Nevertheless, we do not tackle this here and leave it for future research.

The following two results follow similarly as [31, Prop. 2.3] and [31, Prop. 2.4]. We include somewhat alternative proofs for completeness.

Proposition 6.2.6 *If an operator space E has the C-W*CBAP and is locally reflexive, then E has the C-CBAP.*

Proof If E has the C-W*CBAP there is a net of finite-rank maps $\phi_i : E \to E''$ such that $\|\phi_i\|_{\mathrm{cb}} \leq C$ for all i and for every $x \in E$, $\phi_i(x) \to \kappa_E(x)$ in the weak* topology. For each i, let us denote the range of the mapping ϕ_i by $F_i \in \mathrm{OFIN}(E'')$. Since E is locally reflexive for each i, there is a net of complete contractions $(\psi_\eta^i)_\eta \subset \mathrm{CB}(F_i, E)$ approximating the inclusion $F_i \hookrightarrow E''$ in the point-weak* topology. By composing we obtain a net of finite rank mappings $(\psi_\eta^i \circ \phi_i) \subset \mathrm{CB}(E, E)$ which converges in the point-weak topology to the identity $id : E \to E$ satisfying $\left\| \psi_\eta^i \circ \phi_i \right\|_{\mathrm{cb}} \leq C$. A classical convex combinations argument allow us to obtain a net with the same properties but which converges in the point-norm topology to the identity; thus concluding that E has the C-CBAP. $\qquad\square$

Proposition 6.2.7 *If E is an operator space such that E' has the C-W*CBAP, then so does E.*

Proof If E' has the C-W*CBAP, by Theorem 6.2.4 (a) (applied to $F = E$) we obtain that the map $E' \otimes_{\mathrm{proj}} E \to (E'' \otimes_{\min} E')'$ is a complete C-isomorphism with its image. Since both proj and min are symmetric we can flip the spaces (by transposition) to obtain that $E \otimes_{\mathrm{proj}} E' \to (E' \otimes_{\mathrm{min}} E'')'$ is a complete C-isomorphism with its image. This is exactly what is stated in Theorem 6.2.4 (f). Hence, E has the C-W*CBAP. $\qquad \square$

As a corollary we see that under local reflexivity the C-CBAP can be transferred from the dual E' to E.

Corollary 6.2.8 *Let E be an operator space which is locally reflexive. If E' has the C-CBAP, then so does E (in fact, if E' has the C-W*CBAP then E has C-CBAP).*

As a consequence of Theorem 6.2.4 we obtain the following analogue of [23, Thm. 16.2]:

Theorem 6.2.9 *Let E be an operator space and $C \geq 1$. The following are equivalent:*

*(a) E has the C-W*CBAP.*
(b) For every operator space F (or only $F = E'$), the identity $E \otimes_{\overleftarrow{\mathrm{proj}}} F \to E \otimes_{\mathrm{proj}} F$ has cb-norm at most C.
(c) For every operator space F, the identity $E \otimes_{\overleftarrow{\mathrm{proj}}} F \to E \otimes_{\mathrm{proj}} F$ has norm at most C.

*In particular, E has the W*CMAP if and only if*

$$\mathrm{proj} = \overleftarrow{\mathrm{proj}} \quad \text{on } E \otimes F$$

for every operator space F (or only $F = E'$).

Proof By the Duality Theorem 5.2.1 we have the complete isometry $E \otimes_{\overleftarrow{\mathrm{proj}}} F \hookrightarrow (E' \otimes_{\mathrm{min}} F')'$. Also, we know that $\overleftarrow{\mathrm{proj}} \leq \mathrm{proj}$. Therefore, $E \otimes_{\overleftarrow{\mathrm{proj}}} F \to E \otimes_{\mathrm{proj}} F$ has cb-norm at most C (resp. norm at most C) if and only if the mapping $E \otimes_{\mathrm{proj}} F \to (E' \otimes_{\mathrm{min}} F')'$ is a complete C-isomorphism (resp. C-isomorphism) onto its image. Now, Theorem 6.2.4 gives the result. $\qquad \square$

Now we present an operator space version of [23, Lem. 16.2] which enables us to prove the subsequent corollary.

Lemma 6.2.10 *Let α be an o.s. tensor norm, E and F be normed operator spaces, and $n \in \mathbb{N}$. Then $\alpha_n \leq C \overleftarrow{\alpha}_n$ on $E \otimes F$ if and only if*

$$B_{M_n\left((E \otimes_\alpha F)'\right)} \subset C \cdot \overline{B_{M_n(E' \otimes_{\alpha'} F')}}^{\tau_w} \subset M_n\left(\mathrm{CB}(E, F')\right) = \mathrm{CB}\left(E, M_n(F')\right).$$

Proof Let $z \in M_n(E \otimes F)$. By the Duality Theorem 5.2.1,

$$\overleftarrow{\alpha}_n(z; E, F) = \sup\left\{ \|\langle\langle T, z\rangle\rangle\|_{M_{nm}} \mid T \in B_{M_m(E' \otimes_{\alpha'} F')} \right\},$$

whereas clearly

$$\alpha_n(z; E, F) = \sup\left\{ \|\langle\langle T, z\rangle\rangle\|_{M_{nm}} \mid T \in B_{M_m((E \otimes_\alpha F)')} \right\}.$$

By the bipolar theorem, the inclusion

$$B_{M_n\left((E\otimes_\alpha F)'\right)} \subset C \cdot \overline{B_{M_n(E'\otimes_{\alpha'} F')}}^{\tau_w}$$

is then equivalent to $\alpha_n \leq C\overleftarrow{\alpha}_n$ on $M_n(E \otimes F)$. $\square$

From Theorem 6.2.9 and Lemma 6.2.10 we derive the following corollary. Compare to Proposition 6.2.3.

Corollary 6.2.11 *Suppose that E is a normed operator space and $C \geq 1$. The following are equivalent:*

*(a) E has the C-W*CBAP.*
(b) $C \cdot B_{E'\otimes_{\min} E''}$ is τ_w-dense in $B_{\mathrm{CB}(E,E'')}$.
(c) For any operator space F, $C \cdot B_{E'\otimes_{\min} F'}$ is τ_w-dense in $B_{\mathrm{CB}(E,F')}$.

Proof The equivalence between (a) and (c) follows from Theorem 6.2.9 (c) and Lemma 6.2.10. Clearly, (c) implies (b). Assuming (b), any $T \in B_{\mathrm{CB}(E,E'')}$ is the τ_w-limit of a net in $C \cdot B_{E'\otimes_{\min} E''}$, which means T is also the point-weak limit of the same net (by an argument analogous to that of Remark 6.1.1), and hence also the point-weak* limit. Now Proposition 6.2.3 implies (a). $\square$

Chapter 7
Mapping Ideals

We introduce now the definition of a mapping ideal, the operator space version of Banach operator ideals defined by Pietsch and also widely developed by Defant and Floret. The notion in our framework comes from [34, Sec. 12.2], though we have to strengthen Effros and Ruan's definition in order to keep the essence of the traditional Banach space concept and allow the natural relationship with tensor norms. As in [23] the definition of ideal is restricted to the category of complete spaces.

7.1 Definition and Examples

Definition 7.1.1 A *(normed) mapping ideal* $(\mathfrak{A}, \mathbf{A})$ is an assignment, for each pair of operator spaces $E, F \in \mathrm{OBAN}$, of a linear space $\mathfrak{A}(E, F) \subseteq \mathrm{CB}(E, F)$ together with an operator space structure $\mathbf{A}$ on $\mathfrak{A}(E, F)$ such that

(a) The identity map $\mathfrak{A}(E, F) \to \mathrm{CB}(E, F)$ is a complete contraction.
(b) For every $x' \in M_n(E')$ and $y \in M_m(F)$ the mapping $x' \otimes y$ belongs to $M_{nm}(\mathfrak{A}(E, F))$ and $\mathbf{A}_{nm}(x' \otimes y) = \|x'\|_{M_n(E')} \|y\|_{M_m(F)}$.
(c) The ideal property: whenever $T \in M_n(\mathfrak{A}(E, F))$, $r \in \mathrm{CB}(E_0, E)$ and $s \in \mathrm{CB}(F, F_0)$, it follows that $s_n \circ T \circ r$ belongs to $M_n(\mathfrak{A}(E_0, F_0))$ with

$$\mathbf{A}_n(s_n \circ T \circ r) \leq \|s\|_{\mathrm{cb}} \, \mathbf{A}_n(T) \, \|r\|_{\mathrm{cb}} \, .$$

Remark 7.1.2 Item (b) of the previous definition is the *new* requirement, not considered in [34, Sec. 12.2]. In the Banach space realm, the definition of normed operator ideal includes that the ideal norm of the identity $id_{\mathbb{C}} : \mathbb{C} \to \mathbb{C}$ is one [57, Def. 6.1.1.]. This requisite is in fact the key to relate normed ideals and finitely-generated tensor norms. Unfortunately there is no immediate translation of this requirement into the operator space framework, but nevertheless there is a way to

© The Author(s), under exclusive license to Springer Nature Switzerland AG 2025
J. A. Chávez-Domínguez et al., *Operator Space Tensor Norms*, Lecture Notes
in Mathematics 2379, https://doi.org/10.1007/978-3-031-96209-7_7

tackle this issue. By [59, Prop. 6.1.5] the aforementioned condition about the norm of the identity (together with the other properties of normed operator ideals) implies that—and in fact it is equivalent to—the ideal norm of any elementary tensor $x' \otimes y$ (for $x' \in E'$ and $y \in F$) is exactly $\|x'\| \|y\|$. This has inspired item (b) on our definition. Note the similarity between this statement and Eq. (2.1.1) in the definition of o.s. tensor norms.

Let us show now that (b) can be somehow relaxed. First, note that since $\|x' \otimes y\|_{\mathrm{cb}} = \|x'\|_{M_n(E')}\|y\|_{M_m(F)}$, condition (a) implies that $\mathbf{A}_{nm}(x' \otimes y) \geq \|x'\|_{M_n(E')}\|y\|_{M_m(F)}$. Hence, we can replace in (b) the symbol $=$ by $\leq$.

Also, it is enough that this inequality is satisfied for every $E, F \in \mathrm{OFIN}$. Indeed, let us show that the inequality is valid for every Banach operator space when it is so for every finite-dimensional operator space. Let $x' = (x'_{ij}) \in M_n(E')$ and $y = (y_{ij}) \in M_m(F)$ with $E, F \in \mathrm{OBAN}$. Consider $L = \bigcap_{i,j} \ker(x'_{ij}) \in \mathrm{OCOFIN}(E)$ and $u' \in M_n((E/L)')$ such that $x' = (q_L^E)'_n(u')$. Since $(q_L^E)'$ is a complete isometry, we have $\|u'\|_{M_n((E/L)')} = \|x'\|_{M_n(E')}$. Choosing $N \in \mathrm{OFIN}(F)$ such that $y \in M_m(N)$ by means of condition (c) of the definition we obtain

$$\mathbf{A}_{nm}(x' \otimes y; E, F) \leq \mathbf{A}_{nm}(u' \otimes y; E/L, N) \leq \|u'\|_{M_n((E/L)')}\|y\|_{M_m(N)}$$

$$= \|x'\|_{M_n(E')}\|y\|_{M_m(F)}.$$

Therefore, we can exchange condition (b) of the previous definition by the following:

(b') For every $E, F \in \mathrm{OFIN}$, $x' \in M_n(E')$ and $y \in M_m(F)$ the mapping $x' \otimes y$ belongs to $M_{nm}(\mathfrak{A}(E, F))$ and $\mathbf{A}_{nm}(x' \otimes y) \leq \|x'\|_{M_n(E')}\|y\|_{M_m(F)}$.

Of course, the class of completely bounded mappings CB is obviously a mapping ideal.

Let us recall the definitions of some other mapping ideals that have already appeared in the literature. We explain in each case (whenever it is not obvious) why condition (b') is satisfied. Let E and F be Banach operator spaces.

(i) *Finite rank mappings $\mathcal{F}$*: the space $\mathcal{F}(E, F)$ consists of finite rank mappings from E to F. This is a mapping ideal with the cb-norm.

(ii) *Completely approximable mappings $\mathcal{A}$*: the space $\mathcal{A}(E, F)$ is the closure of $\mathcal{F}(E, F)$ inside $\mathrm{CB}(E, F)$.

(iii) *Completely nuclear mappings $\mathcal{N}$* [34, Sec. 12.2]: the space $\mathcal{N}(E, F)$ is the image of the mapping

$$\Phi : E' \widehat{\otimes}_{\mathrm{proj}} F \to E' \widehat{\otimes}_{\mathrm{min}} F \subset \mathrm{CB}(E, F) \tag{7.1.1}$$

with the quotient operator space structure given by the canonical identification

$$\mathcal{N}(E, F) = \frac{E' \widehat{\otimes}_{\mathrm{proj}} F}{\ker \Phi}.$$

The nuclear operator space norm is denoted by ν . Equation (2.1.1) for the proj norm shows that (b') is satisfied for $\mathcal{N}$.

(iv) *Completely integral mappings* $\mathcal{I}$ [34, Sec. 12.3]: the space $\mathcal{I}(E, F)$ is formed by those mappings such that

$$\iota(T) = \sup\{(\nu(T|_M) : M \in \mathrm{OFIN}(E)\} < \infty.$$

Given $T = (T_{ij}) \in M_n(\mathcal{I}(E, F))$ it is defined

$$\iota_n(T) = \sup\{(\nu_n(T|_M) : M \in \mathrm{OFIN}(E)\}.$$

This gives an operator space structure on $\mathcal{I}(E, F)$. Since $\mathcal{I}$ and $\mathcal{N}$ coincide on OFIN, condition (b') is satisfied for $\mathcal{I}$.

(v) *Completely p-summing* Π_p [64, Chap. 5]: For $1 \leq p < \infty$, a linear map $T : E \to F$ belongs to $\Pi_p(E, F)$ if the mapping

$$i_{S_p} \otimes T : S_p \otimes_{\min} E \to S_p[F]$$

is bounded (equivalently, completely bounded), and we denote its norm by $\pi_p(T)$. The operator space structure of $\Pi_p(E, F)$ is inherited from $\mathrm{CB}(S_p \otimes_{\min} E, S_p[F])$. As a consequence of [18, Thm. 3.1] for any $E, F \in$ OFIN we have the following complete isometries:

$$\Pi_p(E, F) = (E \otimes_{d_{p'}} F')' = E' \otimes_{(d_{p'})'} F. \tag{7.1.2}$$

Note that Eq. (2.1.1) for the $(d_{p'})'$ norm shows that (b') is satisfied for Π_p.

(vi) *Exactly integral mappings* $\mathcal{I}^{\mathrm{ex}}$ [34, Chap. 15]: A linear map $T : E \to F$ belongs to $\mathcal{I}^{\mathrm{ex}}(E, F)$ if

$$\iota^{\mathrm{ex}}(T) = \sup\{\left\| i_{G'} \otimes T : G' \otimes_{\min} E \to G' \otimes_{\mathrm{proj}} F \right\| :$$
$$G \subseteq M_n \text{ subspace}, n \in \mathbb{N}\} < \infty. \tag{7.1.3}$$

The reader should be warned that this is not how exactly integral mappings were originally defined in [37] or initially presented in [34, Chap. 15], but this is a characterization in [34, Thm. 15.4.1] that is simpler to state.

It is mentioned in passing in [34, p. 285] that $\mathcal{I}^{\mathrm{ex}}(E, F)$ has a natural operator space structure, but the details are left to the reader. Let us explicitly state one of the various ways to characterize it: for $T = (T_{ij}) \in M_m(\mathcal{I}^{\mathrm{ex}}(E, F))$, we have

$$\iota_m^{\mathrm{ex}}(T) = \sup\{\left\| i_{G'} \otimes T : G' \otimes_{\min} E \to M_m(G' \otimes_{\mathrm{proj}} F) \right\| :$$
$$G \subseteq M_n \text{ subspace}, n \in \mathbb{N}\}. \tag{7.1.4}$$

From this expression and the uniformity of the injective and projective tensor products of operator spaces, it is clear that $\mathcal{I}^{\mathrm{ex}}$ satisfies the ideal property (c).

Comparing to an analogous expression for completely integral mappings [34, Lem. 12.3.1], where the G' above is replaced by an arbitrary finite-dimensional operator space, we immediately conclude that the formal identity $\mathcal{I}(E, F) \to \mathcal{I}^{\mathrm{ex}}(E, F)$ is a complete contraction.

It is also straightforward to check that the formal identity $\mathcal{I}^{\mathrm{ex}}(E, F) \to \Pi_1(E, F)$ is not only a contraction [34, Sec. 15.5], but in fact a complete contraction, which allows us to conclude that conditions (a) and (b) are satisfied.

(vii) *Completely right p-nuclear* $\mathcal{N}^p$ [19, Def. 2.1]: We say that a linear mapping $T : E \to F$ belongs to $\mathcal{N}^p(E, F)$ if it is in the range of the canonical inclusion

$$J^p : E' \widehat{\otimes}_{d_p} F \to E' \widehat{\otimes}_{\min} F \subset \mathrm{CB}(E, F).$$

We endow $\mathcal{N}^p(E, F)$ with the operator space quotient structure $E' \widehat{\otimes}_{d_p} F / \ker J^p$. Equation (2.1.1) for the d_p norm shows that (b') is satisfied for $\mathcal{N}^p$.

(viii) *Operator p-compact mappings* $\mathcal{K}_p$ [19, Def. 3.2]: A mapping $T \in \mathcal{K}_p(E, F)$ if there exist $G \in \mathrm{OBAN}$, a completely right p-nuclear mapping $\Theta \in \mathcal{N}^p(G, F)$ and a completely bounded mapping $R \in \mathrm{CB}(E, G/\ker \Theta)$ with $\|R\|_{\mathrm{cb}} \le 1$ such that the following diagram commutes

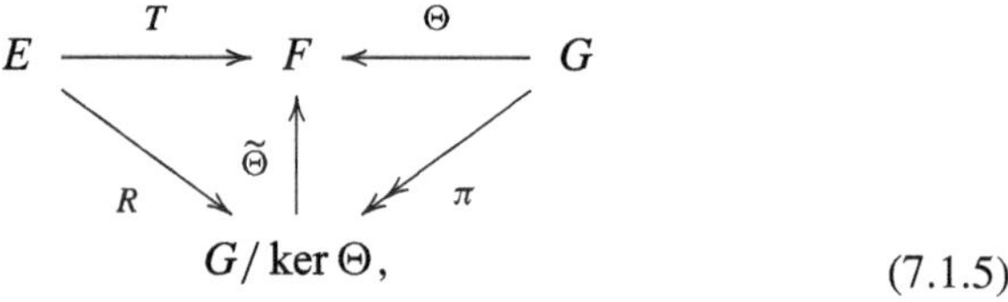

$$\tag{7.1.5}$$

where π and $\widetilde{\Theta}$ stand for the natural quotient mapping and the natural monomorphism associated to Θ respectively.

For $T \in \mathcal{K}_p(E, F)$, we also define

$$\kappa_p(T) := \inf\{\nu^p(\Theta)\},$$

where the infimum runs over all possible completely right p-nuclear mappings $\Theta \in \mathcal{N}^p(G, F)$ as in (7.1.5). It should be mentioned [19, Prop. 3.8.] that $\mathcal{K}_p$ is exactly the surjective hull of $\mathcal{N}^p$ (see Proposition 7.2.8 below). This provides the operator space structure of $\mathcal{K}_p$ and shows that condition (b') is satisfied for this class.

(ix) *Completely p-nuclear mappings* $\mathcal{N}_p$ [44]: A linear map $u : E \to F$ is *completely p-nuclear* (denoted $u \in \mathcal{N}_p(E, F)$) if there exists a factorization of u as

$$ E \xrightarrow{\quad \alpha \quad} S_\infty \xrightarrow{\quad M(a,b) \quad} S_p \xrightarrow{\quad \beta \quad} F $$

where $a, b \in S_{2p}$, $M(a, b)$ is the multiplication operator $x \mapsto axb$, and α, β completely bounded maps. The completely p-nuclear norm of u is defined as

$$ \nu_p(u) = \inf \left\{ \|\alpha\|_{\mathrm{cb}} \|a\|_{S_{2p}} \|b\|_{S_{2p}} \|\beta\|_{\mathrm{cb}} \right\} $$

where the infimum is taken over all factorizations of u as above. From [44, Cor. 3.1.3.9], when $E, F \in \mathrm{OBAN}$ and $1 \le p < \infty$, trace duality yields an isometric isomorphism between $\mathcal{N}_p(E, F)'$ and $\Pi_{p'}(F, E'')$. This allows us to endow $\mathcal{N}_p(E, F)$ with an operator space structure. Also, condition (b') holds since for $E, F \in \mathrm{OFIN}$, $\mathcal{N}_p(E, F) = F \otimes_{d_p} E'$.

(x) *Completely (q, p)-mixing mappings* $\mathfrak{M}_{q,p}$ [17, 77]: Let $1 \le p \le q < \infty$. A mapping $u : E \to F$ is said to be *completely (q, p)-mixing* (denoted $u \in \mathfrak{M}_{q,p}(E, F)$) if there exists a constant K such that for any $G \in \mathrm{OBAN}$ and any completely q-summing map $v : F \to G$, the composition $v \circ u$ is a completely p-summing map and $\pi_p(v \circ u) \le K \pi_q(v)$. The *completely (q, p)-mixing norm* of u is the smallest such K and is denoted by $\mathfrak{m}_{q,p}(u)$. By [17, Thm. 4.3], when $F \subseteq \mathcal{B}(H)$ we have that $u \in \mathfrak{M}_{q,p}(E, F)$ if and only if for all n and all (x_{ij}) in $M_n(E)$ we have

$$ \sup \left\{ \left\| \left(a(ux_{ij})b\right) \right\|_{S_p^n[S_q(H)]} : a, b \in B_{S_{2q}(H)}, \ a, b \ge 0 \right\} $$

$$ \le C \left\| (x_{ij}) \right\|_{S_p^n \otimes_{\min} E}, $$

so in this way $\mathfrak{M}_{q,p}(E, F)$ inherits an operator space structure from $\mathrm{CB}(S_p \otimes_{\min} F, \ell_\infty(S_p[S_q(H)]))$. To verify condition (b'), suppose $E, F \in \mathrm{OFIN}$, $x' \in M_n(E')$ and $y \in M_m(F)$. Let $(x_{ij}) \in M_k(E)$, and consider $a, b \in B_{S_{2q}(H)}$. Observe that for each i, j we have $a\left((x' \otimes y)x_{ij}\right)b = (x' \otimes ayb)(x_{ij})$. Now, since we have already observed that condition (b') is satisfied for the completely p-summing maps, the norm of $x' \otimes ayb$ in $M_{nm}(\mathrm{CB}(S_p \otimes_{\min} E, S_p[S_q(H)]))$ is at most $\|x'\|_{M_n(E')} \|ayb\|_{M_m(S_q(H))}$. Since the multiplication map $w \mapsto awb$ from $\mathcal{B}(H)$ to $S_q(H)$ has cb-norm at most $\|a\|_{S_{2q}} \|b\|_{S_{2q}}$ [64, Prop. 5.6], taking the supremum over a, b yields condition (b') for $\mathfrak{M}_{q,p}$.

(xi) *Mappings factoring through column (resp. row) spaces* Γ_c *(resp. Γ_r)* [32]: $\Gamma_c(E, F)$ denotes the space of linear maps $T : E \to F$ admitting a factorization of the form

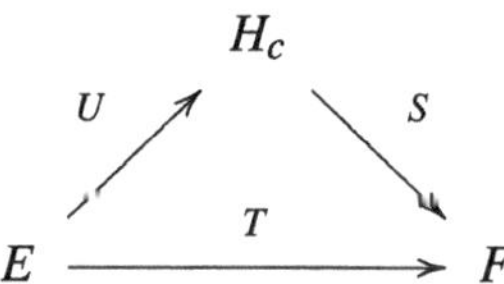

where H_c is a column Hilbert space (that is, a Hilbert space H with the operator space structure given by the identification with $B(\mathbb{C}, H)$) and U, S are completely bounded, endowed with the norm $\gamma_2^c(T) = \inf\{\|U\|_{\mathrm{cb}} \|S\|_{\mathrm{cb}}\}$. More generally, the operator space structure on $\Gamma_c(E, F)$ is defined by setting, for $T \in M_n(\Gamma_c(E, F))$, $(\gamma_2^c)_n(T) = \inf\{\|U\|_{\mathrm{cb}} \|S\|_{\mathrm{cb}}\}$ where the infimum is now taken over all factorizations of the form

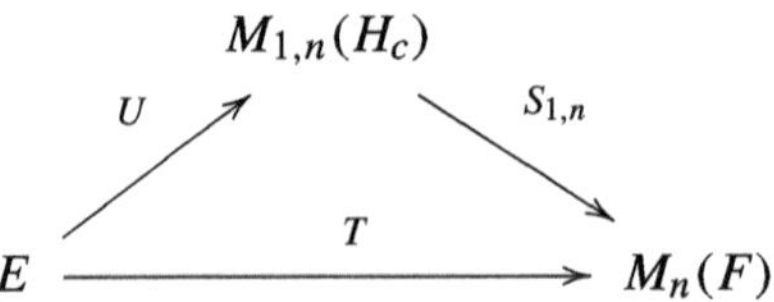

where $U : E \to M_{1,n}(H_c)$ and $S : H_c \to M_{n,1}(F)$ are completely bounded. It is immediate from the definition that properties (a) and (c) are satisfied. To verify (b'), let $x' \in M_n(E')$ and $y \in M_m(F)$. Recall that in the case $H = \ell_2^n$, it is traditional to denote the corresponding column Hilbert space by C_n. From the identifications $M_n(E') = \mathrm{CB}(E, M_n)$ and $M_n = M_{1,n}(C_n)$, we can identify x' with an operator $U \in \mathrm{CB}(E, M_{1,n}(C_n))$ satisfying $\|U\|_{\mathrm{cb}} = \|x'\|_{M_n(E')}$. Consider now the linear map $\widehat{y} : \mathbb{C} \to M_m(F)$ which sends 1 to y, which has rank at most one, hence it is completely bounded and has norm $\|y\|_{M_m(F)}$. Let $S = \widehat{y}_{n,1} : M_{n,1}(\mathbb{C}) \to M_{nm,1}(F)$, which also has cb-norm equal to $\|y\|_{M_m(F)}$. Since $M_{n,1}(\mathbb{C}) = C_n$, note that this yields the desired factorization $x' \otimes y = S_{1,n} \circ U$ with $\|U\|_{\mathrm{cb}} \|S\|_{\mathrm{cb}} \leq \|x\|_{M_n(E')} \|y\|_{M_m(F)}$.

Analogously, one defines the space $\Gamma_r(E, F)$ as those linear maps $T : E \to F$ admitting a factorization of the form

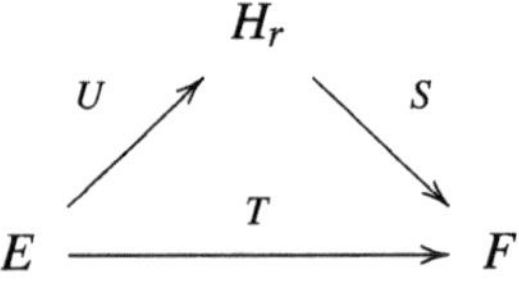

where H_r is a row Hilbert space (that is, a Hilbert space H with the operator space structure given by the identification $H_r = \mathcal{B}(H', \mathbb{C})$) and U, S are completely bounded. The operator space structure is defined by setting, for $T \in M_n(\Gamma_r(E, F))$, $(\gamma_2^r)_n(T) = \inf\{\|U\|_{\mathrm{cb}} \|S\|_{\mathrm{cb}}\}$ where the infimum is now taken over all factorizations of the form

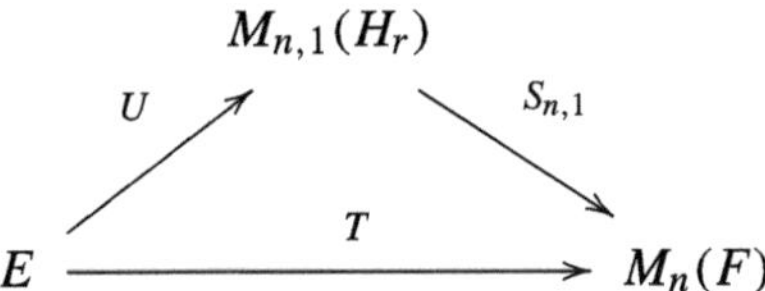

where $U : E \to M_{n,1}(H_r)$ and $S : H_r \to M_{1,n}(F)$ are completely bounded. Properties (a), (b'), and (c) follow analogously to the column case.

(xii) *Completely compact mappings* [74]: A linear mapping $T \in CB(E, F)$ is said to be *completely compact* (denoted $T \in CC(E, F)$) if for every $\varepsilon > 0$ there exists a finite-dimensional $F_\varepsilon \subseteq F$ such that for every $x \in M_n(E)$ with $\|x\|_{M_n(E)} \le 1$ there exists $y \in M_n(F_\varepsilon)$ such that $\|T_n x - y\|_{M_n(F)} \le \varepsilon$. The completely compact mappings form a closed subspace of $CB(E, F)$, inheriting the corresponding operator space structure, and it is easy to check that they satisfy the ideal property.

Naturally, the list above is not meant to be exhaustive. Without going into the details, we point out that [44] defines a number of other potential examples of mapping ideals including: completely p-integral mappings (Definition 3.1.3.2) the completely injective hull of the completely p-nuclear mappings (Definition 3.1.3.5), completely (p, k)-mappings and (p, k)-integral maps (Definition 3.1.3.6), mappings admitting completely bounded factorizations through diagonal operators from ℓ_p to ℓ_q and their noncommutative analogue, namely two-sided multiplication operators from S_p to S_q (Sect. 3.1.4). While initially only a norm is defined, characterizations are proved that allow us to obtain o.s. structures. Observe that by Proposition 7.2.3, the aforementioned completely injective hull of the completely p-nuclear mappings is indeed a mapping ideal. Note that this is an operator space version of the classical ideal of quasi p-nuclear operators (often denoted $\mathcal{QN}_p$) introduced in [56] and further studied in e.g. [25, 70]. For all the other aforementioned examples from [44] we have not verified that condition (b') is satisfied, but we expect that to be the case.

The following definition concerns the completeness of $\mathfrak{A}(E, F)$.

Definition 7.1.3 A mapping ideal $\mathfrak{A}$ is called a *Banach mapping ideal* if $\mathfrak{A}(E, F)$ is a Banach space, for every $E, F \in OBAN$.

Note that all the examples listed above are Banach mapping ideals, except for the ideal of finite rank mappings.

As in the Banach space setting, each o.s. tensor norm allows us to construct a mapping ideal "dual" to it. For the reader familiar with the theory connecting tensor norms and ideals of mappings in Banach spaces, we stress that this dual construction should not be confused with the classical association between maximal operator ideals and finitely-generated tensor norms, which will be discussed in Chap. 8.

Example 7.1.4 Given an o.s. tensor norm α, for $E, F \in OBAN$ let

$$\mathfrak{A}_\alpha(E, F) := (E \otimes_\alpha F')' \cap CB(E, F).$$

Then, $\mathfrak{A}_\alpha$ is a mapping ideal.

Indeed, the complete contraction $E \otimes_{\mathrm{proj}} F' \hookrightarrow E \otimes_\alpha F'$ implies that the identity map $\mathfrak{A}_\alpha(E, F) \subset CB(E, F)$ is completely contractive too. Also, for $E, F \in OFIN$, $\mathfrak{A}_\alpha(E, F) = E' \otimes_{\alpha'} F$ and therefore condition (b') of the definition holds. On the other hand, the metric mapping property of α readily implies the ideal property of $\mathfrak{A}_\alpha$. The subscript α in the notation indicates the α-continuity of the mappings. It is important to emphasize that, according to Definition 8.2.1 below, we have $\mathfrak{A}_\alpha \sim \alpha'$.

7.2 Mapping Procedures

We first consider ideals that "do not change when followed by a complete isometry".
To be precise:

Definition 7.2.1 A mapping ideal $\mathfrak{A}$ is said to be *completely injective* if for each complete isometry $i : F \to G$ and $T \in M_n(\mathrm{CB}(E, F))$ the following equivalence holds: $T \in M_n(\mathfrak{A}(E, F))$ if and only if $i_n \circ T \in M_n(\mathfrak{A}(E, G))$, with $\mathbf{A}_n(T) = \mathbf{A}_n(i_n \circ T)$.

It straightforward that the mapping ideal CB is completely injective. As a consequence we also have that Π_p is completely injective. Indeed, let $T \in M_n(\mathrm{CB}(E, F))$ and $i : F \to G$ a complete isometry such that $i_n \circ T \in M_n(\Pi_p(E, G))$. Since

$$M_n(\Pi_p(E, G)) \hookrightarrow M_n(\mathrm{CB}(S_p \otimes_{\min} E, S_p[G]))$$

and i induces a complete isometry from $S_p[F]$ to $S_p[G]$ by Theorem 1.4.3, we obtain that $T \in M_n(\Pi_p(E, F))$.

Remark 7.2.2 Observe that to prove that a mapping ideal $\mathfrak{A}$ is completely injective it is enough to check the following two conditions:

(a) for each complete isometry $i : F \to G$ and $T \in \mathrm{CB}(E, F)$ such that $i \circ T \in \mathfrak{A}(E, G)$ we have that $T \in \mathfrak{A}(E, F)$.
(b) $\mathbf{A}_n(T) \leq \mathbf{A}_n(i_n \circ T)$ for every complete isometry $i : F \to G$ and $T \in M_n(\mathfrak{A}(E, F))$.

A normed operator space E has the *complete C-extension property* if for every $T \in \mathrm{CB}(E, F)$ and a complete injection $i : E \to G$ there is $\widetilde{T} \in \mathrm{CB}(G, F)$ such that the following diagram commutes

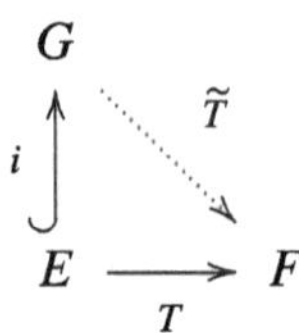

with $\|\widetilde{T}\|_{\mathrm{cb}} \leq C \|T\|_{\mathrm{cb}}$.
Recall that $\mathcal{B}(H)$ has the complete 1-extension property [65, Thm. 1.6].

Given a mapping ideal $(\mathfrak{A}, \mathbf{A})$, the following proposition allows us to talk about the smallest completely injective ideal that contains it.

Proposition 7.2.3 *Let $(\mathfrak{A}, \mathbf{A})$ be a mapping ideal. Then there is a (unique) smallest completely injective mapping ideal $\mathfrak{A}^{\mathrm{inj}}$ which contains $\mathfrak{A}$ (called the* completely *injective hull of $\mathfrak{A}$). Moreover, if V is a completely injective operator space and $i : F \to V$ is a complete isometry, then $T \in M_n(\mathrm{CB}(E, F))$ is in $M_n(\mathfrak{A}^{\mathrm{inj}}(E, F))$ if and only if $i_n \circ T \in M_n(\mathfrak{A}(E, V))$ with*

$$\mathbf{A}_n^{\mathrm{inj}}(T) = \mathbf{A}_n(i_n \circ T).$$

Proof Given $F \in \mathrm{OBAN}$ we denote by $i_F : F \hookrightarrow \mathcal{B}(H_F)$ any embedding that provides F the operator space structure. We define

$$\mathfrak{A}^{\mathrm{inj}}(E, F) := \{T \in \mathrm{CB}(E, F) : i_F \circ T \in \mathfrak{A}(E, \mathcal{B}(H_F))\},$$

endowed with the operator space norm defined for $T \in M_n(\mathrm{CB}(E, F))$ as

$$\mathbf{A}_n^{\mathrm{inj}}(T) := \mathbf{A}_n((i_F)_n \circ T).$$

To see that this is well defined, let us prove that if $i : F \to V$ is any complete isometry of F into a completely injective space V (which has to be complete as explained in Sect. 1.1.3) then for every $n \in \mathbb{N}$ and $T \in M_n(\mathrm{CB}(E, F))$ such that $i_n \circ T \in M_n(\mathfrak{A}(E, V))$ we have the equality

$$\mathbf{A}_n(i_n \circ T) = \mathbf{A}_n((i_F)_n \circ T). \tag{7.2.1}$$

Indeed, by the injectivity of $\mathcal{B}(H_F)$ there is a mapping $\widetilde{i}_F : V \to \mathcal{B}(H_F)$ such that $i_F = \widetilde{i}_F \circ i$ with $\|\widetilde{i}_F\|_{\mathrm{cb}} = 1$. Then, by the ideal property

$$\mathbf{A}_n((i_F)_n \circ T) = \mathbf{A}_n((\widetilde{i}_F)_n \circ i_n \circ T) \leq \mathbf{A}_n(i_n \circ T).$$

On the other hand, by the injectivity of V, we can argue as before to show the existence of a mapping $\widetilde{i} : \mathcal{B}(H_F) \to V$ such that $i = \widetilde{i} \circ i_F$ with $\|\widetilde{i}\|_{\mathrm{cb}} = 1$ and therefore

$$\mathbf{A}_n(i_n \circ T) = \mathbf{A}_n((\widetilde{i})_n \circ (i_F)_n \circ T) \leq \mathbf{A}_n((i_F)_n \circ T).$$

This shows Eq. (7.2.1).

To see that $\mathfrak{A}^{\mathrm{inj}}$ is a mapping ideal, we will only prove the ideal property since the other conditions easily follow. Let $T \in M_n(\mathfrak{A}^{\mathrm{inj}}(E, F))$, $r \in \mathrm{CB}(E_0, E)$ and $s \in \mathrm{CB}(F, F_0)$. Then,

$$\mathbf{A}_n^{\mathrm{inj}}(s_n \circ T \circ r) = \mathbf{A}_n((i_{F_0})_n \circ s_n \circ T \circ r) \leq \mathbf{A}_n((i_{F_0} \circ s)_n \circ T)\|r\|_{\mathrm{cb}}$$

Once again the injectivity of $\mathcal{B}(H_{F_0})$ provides us with a mapping $\widetilde{s} \in \mathrm{CB}(\mathcal{B}(H_F), \mathcal{B}(H_{F_0}))$ satisfying $i_{F_0} \circ s = \widetilde{s} \circ i_F$, with $\|\widetilde{s}\|_{\mathrm{cb}} \leq \|s\|_{\mathrm{cb}}$. Thus,

$$\mathbf{A}_n^{\mathrm{inj}}(s_n \circ T \circ r) \leq \mathbf{A}_n((\widetilde{s} \circ i_F)_n \circ T)\|r\|_{\mathrm{cb}} \leq \|s\|_{\mathrm{cb}}\mathbf{A}_n((i_F)_n \circ T)\|r\|_{\mathrm{cb}}$$

$$= \|s\|_{\mathrm{cb}}\mathbf{A}_n^{\mathrm{inj}}(T)\|r\|_{\mathrm{cb}}.$$

We now see that $\mathfrak{A}^{\mathrm{inj}}$ is completely injective: let $i : F \to G$ be a complete isometry, let $T \in \mathrm{CB}(E, F)$ and suppose that $i \circ T \in \mathfrak{A}^{\mathrm{inj}}(E, G)$. If $i_G : G \hookrightarrow \mathcal{B}(H_G)$ is a completely isometric embedding, the extension property of $\mathcal{B}(H_G)$ shows the existence of a mapping $\widetilde{i_F}$ such that

$$\widetilde{i_F} \circ (i_G \circ i) = i_F. \tag{7.2.2}$$

Now, the definition of the ideal implies $i_G \circ i \circ T \in \mathfrak{A}(E, \mathcal{B}(H_G))$. Then, $\widetilde{i_F} \circ (i_G \circ i) \circ T = i_F \circ T$ belongs to $\mathfrak{A}(E, \mathcal{B}(H_F))$. Hence, $T \in \mathfrak{A}^{\mathrm{inj}}(E, F)$.

Now, if $T \in M_n(\mathfrak{A}^{\mathrm{inj}}(E, F))$ and $i : F \to G$ is a complete isometry, we have

$$\mathbf{A}_n^{\mathrm{inj}}(T) = \mathbf{A}_n((i_F)_n \circ T) = \mathbf{A}_n((\widetilde{i_F} \circ i_G \circ i)_n \circ T)$$

$$\leq \mathbf{A}_n((i_G)_n \circ (i_n \circ T)) = \mathbf{A}_n^{\mathrm{inj}}(i_n \circ T).$$

That $\mathfrak{A}^{\mathrm{inj}}$ is the smallest completely injective ideal which contains $\mathfrak{A}$ is trivial by definition of $\mathfrak{A}^{\mathrm{inj}}$. The last part of the statement follows from what was done at the beginning of the proof to show that $\mathbf{A}_n^{\mathrm{inj}}(T)$ was well defined. $\qquad\square$

Example 7.2.4 For the mapping ideal $\mathcal{I}$ of completely integral mappings, the previous proposition says that $T \in \mathcal{I}^{\mathrm{inj}}(E, F)$ if and only if $i_F \circ T \in \mathcal{I}(E, \mathcal{B}(H_F))$. By [34, Prop. 15.5.1], $\mathcal{I}(E, \mathcal{B}(H_F)) = \Pi_1(E, \mathcal{B}(H_F))$. Since we have already explained that Π_1 is a completely injective mapping ideal we derive that $\mathcal{I}^{\mathrm{inj}} = \Pi_1$ (as it happens in the Banach space framework).

The following simple result shows that the inj-procedure respects completeness.

Proposition 7.2.5 *Let* $(\mathfrak{A}, \mathbf{A})$ *be a Banach mapping ideal then* $(\mathfrak{A}^{\mathrm{inj}}, \mathbf{A}^{\mathrm{inj}})$ *is also a Banach mapping ideal.*

Proof Let $E, F \in \mathrm{OBAN}$, we only have to check that $\mathfrak{A}^{\mathrm{inj}}(E, F)$ is complete. Let $(T_k)_{k \in \mathbb{N}}$ be a Cauchy sequence in $\mathfrak{A}^{\mathrm{inj}}(E, F)$. Note that $(T_k)_{k \in \mathbb{N}}$ is also a Cauchy sequence in $\mathrm{CB}(E, F)$ and then it converges (in the cb-norm) to an operator $T \in \mathrm{CB}(E, F)$. On the other hand, $(i_F \circ T_k)_{k \in \mathbb{N}}$ is a Cauchy sequence in $\mathfrak{A}(E, \mathcal{B}(H_F))$ which converges to an operator $S \in \mathfrak{A}(E, \mathcal{B}(H_F))$. To finish, it remains to prove that $S = i_F \circ T$. Indeed, note that $T_k \to T$ pointwise and therefore $i_F \circ T_k \to i_F \circ T$ pointwise as well. Now the result follows by the uniqueness of the limit.

$\qquad\square$

As a corollary we have an analogous version of [23, Cor. 9.7] for the operator space setting.

Corollary 7.2.6 *Let* $F \in$ OBAN *with the complete C-extension property and* $(\mathfrak{A}, \mathbf{A})$ *be a mapping ideal. Then for every* $n \in \mathbb{N}$ *and every* $E \in$ OBAN *we have* $M_n(\mathfrak{A}(E, F)) = M_n(\mathfrak{A}^{\mathrm{inj}}(E, F))$. *Moreover,*

$$\mathbf{A}_n(T) \leq C \mathbf{A}_n^{\mathrm{inj}}(T),$$

for all $T \in M_n(\mathfrak{A}^{\mathrm{inj}}(E, F))$.

Similarly, we can consider ideals that behave well when composed with complete metric surjections.

Definition 7.2.7 A mapping ideal $(\mathfrak{A}, \mathbf{A})$ is said to be *completely surjective* if for each complete metric surjection $q : G \twoheadrightarrow E$ and $T \in M_n(\mathrm{CB}(E, F))$ the following equivalence holds: $T \in M_n(\mathfrak{A}(E, F))$ if and only if $T \circ q \in M_n(\mathfrak{A}(G, F))$, with $\mathbf{A}_n(T) = \mathbf{A}_n(T \circ q)$.

It is immediate to see that the mapping ideal CB is completely surjective.

Given a mapping ideal $(\mathfrak{A}, \mathbf{A})$, the following proposition defines the smallest completely surjective ideal that contains it. This was already proved in [19, Prop. 3.6], we include a detailed proof for completeness.

Proposition 7.2.8 *Let* $(\mathfrak{A}, \mathbf{A})$ *be a mapping ideal. Then there is a (unique) smallest completely surjective mapping ideal* $\mathfrak{A}^{\mathrm{sur}}$ *which contains* $\mathfrak{A}$ *(called the* completely surjective hull *of* $\mathfrak{A}$*). Moreover: if* G *is a completely projective Banach operator space,* $E \in$ OBAN *and* $q : G \twoheadrightarrow E$ *is a complete metric surjection, then* $T \in M_n(\mathrm{CB}(E, F))$ *is in* $M_n(\mathfrak{A}^{\mathrm{sur}}(E, F))$ *if and only if* $T \circ q \in M_n(\mathfrak{A}(G, F))$ *with*

$$\mathbf{A}_n^{\mathrm{sur}}(T) = \mathbf{A}_n(T \circ q).$$

Proof We define

$$\mathfrak{A}^{\mathrm{sur}}(E, F) = \{T \in \mathrm{CB}(E, F) : T \circ q_E \in \mathfrak{A}(Z_E, F)\},$$

endowed with the operator space structure defined in the statement. First we see that it is well defined. To see this, suppose that G is a completely projective Banach operator space and $q : G \twoheadrightarrow E$ is a complete metric surjection and $T \in M_n(\mathrm{CB}(E, F))$ such that $T \circ q \in M_n(\mathfrak{A}(G, F))$, we have to see that

$$\mathbf{A}_n(T \circ q_E) = \mathbf{A}_n(T \circ q). \tag{7.2.3}$$

By the complete projectivity of Z_E there is a mapping $\widetilde{q}_E : Z_E \to G$ such that $q_E = q \circ \widetilde{q}_E$ with $\|\widetilde{q}_E\|_{\mathrm{cb}} \leq 1$. Then,

$$\mathbf{A}_n(T \circ q_E) = \mathbf{A}_n(T \circ q \circ \widetilde{q}_E) \leq \mathbf{A}_n(T \circ q).$$

On the other hand, by the projectivity of G there is a mapping $\widetilde{q} : G \to Z_E$ such that $q = q_E \circ \widetilde{q}$ with $\|\widetilde{q}\|_{\mathrm{cb}} \leq 1$. As before,

$$\mathbf{A}_n(T \circ q) = \mathbf{A}_n(T \circ q_E \circ \widetilde{q}) \leq \mathbf{A}_n(T \circ q_E).$$

This shows Eq. (7.2.3).

To see that $\mathfrak{A}^{\mathrm{sur}}$ is a mapping ideal, we will only prove the ideal property since the other conditions easily follow.

Let $T \in M_n(\mathfrak{A}^{\mathrm{sur}}(E, F))$, $r \in \mathrm{CB}(E_0, E)$ and $s \in \mathrm{CB}(F, F_0)$. Since Z_{E_0} is completely projective, given $\varepsilon > 0$, there is a lifting $L_\varepsilon \in \mathrm{CB}(Z_{E_0}, Z_E)$ of $r \circ q_{E_0}$ with $\|L_\varepsilon\|_{\mathrm{cb}} \leq (1 + \varepsilon)\|r\|_{\mathrm{cb}}$ such that the following diagram commutes

$$
\begin{array}{ccccccc}
E_0 & \xrightarrow{\ r\ } & E & \xrightarrow{\ T_{ij}\ } & F & \xrightarrow{\ s\ } & F_0 \ .\\
\big\uparrow{\scriptstyle q_{E_0}} & & \big\uparrow{\scriptstyle q_E} & & & & \\
Z_{E_0} & \dashrightarrow & Z_E & & & & \\
 & \scriptstyle L_\varepsilon & & & & &
\end{array}
$$

Then, for every $1 \leq i, j \leq n$ we have $s \circ T_{ij} \circ r \circ q_{E_0} = s \circ T_{ij} \circ q_E \circ L_\varepsilon \in \mathfrak{A}(Z_{E_0}, F_0)$ and hence $s \circ T_{ij} \circ r \in \mathfrak{A}^{\mathrm{sur}}(E_0, F_0)$. Moreover, by the ideal property of $\mathfrak{A}$,

$$\mathbf{A}_n^{\mathrm{sur}}(s_n \circ T \circ r) = \mathbf{A}_n(s_n \circ T \circ r \circ q_{E_0}) = \mathbf{A}_n(s_n \circ T \circ q_E \circ L_\varepsilon)$$

$$\leq \mathbf{A}_n(s_n \circ T \circ q_E)(1 + \varepsilon)\|r\|_{\mathrm{cb}}$$

$$\leq (1 + \varepsilon)\|s\|_{\mathrm{cb}}\mathbf{A}_n^{\mathrm{sur}}(T)\|r\|_{\mathrm{cb}}.$$

We now prove that the mapping ideal $\mathfrak{A}^{\mathrm{sur}}$ is surjective. Let $q : G \twoheadrightarrow E$ be a complete quotient mapping and $T \in \mathrm{CB}(E, F)$ such that $T \circ q \in \mathfrak{A}^{\mathrm{sur}}(G, F)$. Since Z_E is completely projective and $q \circ q_G : Z_G \twoheadrightarrow E$ is a complete quotient mapping, given $\varepsilon > 0$, there is a lifting $L_\varepsilon : Z_E \to Z_G$ of q_E, with cb-norm less than or equal to $1 + \varepsilon$.

We have $T \circ q_E = T \circ q \circ q_G \circ L_\varepsilon \in \mathfrak{A}(Z_E, F)$ since $T \circ q \circ q_G \in \mathfrak{A}(Z_G, F)$. Then, $T \in \mathfrak{A}^{\mathrm{sur}}(E, F)$. Also, if $T \in M_n(\mathfrak{A}^{\mathrm{sur}}(E, F))$ we obtain

$$\mathbf{A}_n^{\mathrm{sur}}(T) = \mathbf{A}_n(T \circ q_E) = \mathbf{A}(T \circ q \circ q_G \circ L_\varepsilon) \leq (1 + \varepsilon)\mathbf{A}_n(T \circ q \circ q_G)$$

$$= (1 + \varepsilon)\mathbf{A}_n^{\mathrm{sur}}(T \circ q) \leq (1 + \varepsilon)\mathbf{A}_n^{\mathrm{sur}}(T).$$

That $\mathfrak{A}^{\mathrm{sur}}$ is the smallest completely projective ideal is trivial by definition of $\mathfrak{A}^{\mathrm{sur}}$. The last part of the statement follows from what was done at the beginning of the proof to show that $\mathbf{A}_n^{\mathrm{sur}}(T)$ was well defined. $\square$

It was proven in [19, Prop. 3.8] that

$$\mathcal{K}_p = (\mathcal{N}^p)^{\mathrm{sur}}.$$

As a consequence, we obviously have that $\mathcal{K}_p$ is a surjective mapping ideal.

As a corollary of the previous proposition we have an analogous version of [23, Cor. 9.8] for the operator space setting.

Corollary 7.2.9 *Let E be a completely projective Banach operator space and $(\mathfrak{A}, \mathbf{A})$ be a mapping ideal. Then for every $F \in$ OBAN we have that $\mathfrak{A}(E, F)$ and $\mathfrak{A}^{\mathrm{sur}}(E, F)$ are completely isometric.*

As for the inj-procedure, we have that the sur-procedure also respects completeness. The proof is analogous to that of Proposition 7.2.5 so we omit it.

Proposition 7.2.10 *Let $(\mathfrak{A}, \mathbf{A})$ be a Banach mapping ideal then $(\mathfrak{A}^{\mathrm{sur}}, \mathbf{A}^{\mathrm{sur}})$ is also a Banach mapping ideal.*

We now define the dual procedure of a mapping ideal.

Definition 7.2.11 If $(\mathfrak{A}, \mathbf{A})$ is a mapping ideal, define for $E, F \in$ OBAN

$$\mathfrak{A}^{\mathrm{dual}}(E, F) := \{T \in \mathrm{CB}(E, F) \,:\, T' \in \mathfrak{A}(F', E')\}$$

and

$$\mathbf{A}_n^{\mathrm{dual}}(T) := \mathbf{A}_n(T'), \quad T \in M_n(\mathfrak{A}^{\mathrm{dual}}(E, F)).$$

It is readily seen that $(\mathfrak{A}^{\mathrm{dual}}, \mathbf{A}^{\mathrm{dual}})$ is also a mapping ideal.

It is clear from the definition that the ideal of completely approximable mappings satisfies that $\mathfrak{A} = \mathfrak{A}^{\mathrm{dual}}$. On the other hand, by [34, Prop. 12.2.5], if $T \in \mathcal{N}(E, F)$ we have that $T \in \mathcal{N}^{\mathrm{dual}}(E, F)$ with $\nu^{\mathrm{dual}}(T) = \nu(T') \le \nu(T)$. In fact, more is true: the same proof shows that the map $\mathcal{N}(E, F) \to \mathcal{N}(F', E')$ given by $T \mapsto T'$ is a complete contraction, and thus the identity map $\mathcal{N}(E, F) \to \mathcal{N}^{\mathrm{dual}}(E, F)$ is a complete contraction as well. In the case when E' has the operator approximation property the map Φ in (7.1.1) is injective [34, Thm. 11.2.5], and therefore the same proof as in [34, Prop. 12.2.6] shows that the two maps above are in fact complete isometries.

Let us also mention that mappings belonging to the ideals $(\Pi_p)^{\mathrm{dual}}$ have already been considered in the literature. For example, [64, Thm. 6.5] characterizes mappings that admit a completely bounded factorization through Pisier's operator Hilbert space OH in terms of conditions involving both Π_2 and $(\Pi_2)^{\mathrm{dual}}$, and [64, Cor. 7.2.2] similarly characterizes mappings admitting a completely bounded factorization through finite-dimensional Schatten p-spaces with conditions that involve $\Pi_{p'}$ and $(\Pi_p)^{\mathrm{dual}}$.

Chapter 8
Maximal Operator Space Mapping Ideals

In the Banach space setting there is an intrinsic relation between tensor norms and operator ideals. As mentioned before, the same happens in the non-commutative context. In this chapter we develop the theory of maximal mapping ideals, showing that the latter and the theory of o.s. tensor norms are two sides of the same coin.

8.1 The Maximal Hull

If $(\mathfrak{A}, \mathbf{A})$ is a mapping ideal, then

$$M \otimes_\alpha N := \mathfrak{A}(M', N)$$

defines an o.s. tensor norm α on OFIN; in other words, if $z \in M_n(M \otimes N)$ and $T_z \in M_n(\mathrm{CB}(M', N))$ is the associated matrix of linear operators, then

$$\alpha_n(z; M, N) := \mathbf{A}_n(T_z : M' \to N).$$

Since the natural maps $\mathfrak{A}(M', N) \to \mathrm{CB}(M', N)$ and $M \otimes_{\min} N \to \mathrm{CB}(M', N)$ are respectively a complete contraction and a complete isometry, it follows that

$$M \otimes_\alpha N \to M \otimes_{\min} N$$

is a complete contraction.

Note that Eq. (2.1.1) follows immediately from condition (b) in the definition of mapping ideals. Also, the ideal property of $\mathfrak{A}$ implies the complete metric mapping property of α: for any linear maps $r \in \mathrm{CB}(M, M_0)$ and $s \in \mathrm{CB}(N, N_0)$, note that for any $z \in M_n(M \otimes N)$

$$s_n \circ T_z \circ r' = T_{(r \otimes s)(z)}$$

© The Author(s), under exclusive license to Springer Nature Switzerland AG 2025
J. A. Chávez-Domínguez et al., *Operator Space Tensor Norms*, Lecture Notes
in Mathematics 2379, https://doi.org/10.1007/978-3-031-96209-7_8

and thus

$$\alpha_n\big((r \otimes s)(z)\big) = \mathbf{A}_n(T_{(r\otimes s)(z)}) = \mathbf{A}_n(s_n \circ T_z \circ r') \le \|s\|_{\mathrm{cb}}\, \mathbf{A}_n(T_z)\, \|r'\|_{\mathrm{cb}}$$

$$= \|s\|_{\mathrm{cb}}\, \|r\|_{\mathrm{cb}}\, \alpha_n(z).$$

Conversely, let α be an o.s. tensor norm on OFIN satisfying the complete metric mapping property. Define for $M, N \in$ OFIN and $T \in M_n(\mathrm{CB}(M, N))$

$$\mathbf{A}_n(T : M \to N) := \alpha_n(z_T; M', N).$$

where $z_T \in M_n(M' \otimes N)$ is the associated matrix of tensors. Let $r \in \mathrm{CB}(M_0, M)$ and $s \in CB(N, N_0)$. Again, condition (b) is related with Eq. (2.1.1). Note that

$$z_{s_n \circ T \circ r} = \big(r' \otimes s\big) z_T$$

and thus, by the complete metric mapping property of α,

$$\mathbf{A}_n(s_n \circ T \circ r) = \alpha_n(z_{s_n \circ T \circ r}) \le \|r'\|_{\mathrm{cb}}\, \|s\|_{\mathrm{cb}}\, \alpha_n(z_T) = \|r\|_{\mathrm{cb}}\, \|s\|_{\mathrm{cb}}\, \mathbf{A}_n(T).$$

Therefore, in the finite-dimensional setting there is a full correspondence between mapping ideals and o.s. tensor norms. The rest of this chapter is devoted to the study of what can be said in the infinite-dimensional setting. Naturally, the answer is satisfactory for mapping ideals that can be well-approximated by finite-dimensional pieces, the idea that motivates the next definition.

Definition 8.1.1 Assume $(\mathfrak{A}, \mathbf{A})$ is a mapping ideal. For $E, F \in$ OBAN and $T \in M_n(\mathrm{CB}(E, F))$, define

$$\mathbf{A}_n^{\max}(T) = \sup\Big\{\mathbf{A}_n((q_L^F)_n \circ T \circ i_M^E)\colon M \in \mathrm{OFIN}(E), L \in \mathrm{OCOFIN}(F)\Big\}$$

and

$$\mathfrak{A}^{\max}(E, F) = \big\{T \in \mathrm{CB}(E, F) : \mathbf{A}_1^{\max}(T) < \infty\big\}.$$

We call $(\mathfrak{A}^{\max}, \mathbf{A}^{\max})$ the *maximal hull* of $(\mathfrak{A}, \mathbf{A})$, and $(\mathfrak{A}, \mathbf{A})$ is called *maximal* if $(\mathfrak{A}, \mathbf{A}) = (\mathfrak{A}^{\max}, \mathbf{A}^{\max})$.

Proposition 8.1.2 *Let $(\mathfrak{A}, \mathbf{A})$ be a mapping ideal. Then $(\mathfrak{A}^{\max}, \mathbf{A}^{\max})$ is a Banach mapping ideal which, of course, is maximal.*

Proof It is clear that each $\mathbf{A}_n^{\max}$ is a norm, and that $\mathbf{A}^{\max}$ satisfies Ruan's axioms (say, as in [34, Prop. 2.3.6]).

Using that the identity $\mathfrak{A}(E, F) \to \mathrm{CB}(E, F)$ is a complete contraction, and the ideal property for the cb-norm, for $T \in M_n(\mathrm{CB}(E, F))$ we have that

$$\mathbf{A}_n\big((q_L^F)_n \circ T \circ i_M^E\big) \leq \left\| (q_L^F)_n \circ T \circ i_M^E \right\|_{M_n(\mathrm{CB}(M,F/L))} < \|T\|_{M_n(\mathrm{CB}(E,F))},$$

which implies that the identity $\mathfrak{A}^{\max}(E,F) \to \mathrm{CB}(E,F)$ is also a complete contraction. Similarly, the ideal property for $(\mathfrak{A}, \mathbf{A})$ shows that all finite-rank operators belong to $\mathfrak{A}^{\max}$ since they belong to $\mathfrak{A}$. Moreover, given $x' \in M_n(E')$ and $y \in M_m(F)$ we have

$$\mathbf{A}_{nm}\big((q_L^F)_{nm} \circ x' \otimes y \circ i_M^E\big) = \mathbf{A}_{nm}\big(x' \circ i_M^E \otimes (q_L^F)_m(y)\big) = \|x' \circ i_M^E\|_{\mathrm{cb}} \|(q_L^F)_m(y)\|_{\mathrm{cb}},$$

showing that $\mathbf{A}_{nm}^{\max}(x' \otimes y) \leq \|x'\|_{\mathrm{cb}} \|y\|_{\mathrm{cb}}$.

Suppose now that $r \in \mathrm{CB}(E_0, E)$ and $s \in \mathrm{CB}(F, F_0)$. Let $M_0 \in \mathrm{OFIN}(E_0)$ and $L_0 \in \mathrm{OCOFIN}(F_0)$. Define $M = r(M_0) \in \mathrm{OFIN}(E)$ and $L = s^{-1}(L_0) \in \mathrm{OCOFIN}(F)$. Note that $r = i_M^E \circ r$, and $q_{L_0}^{F_0} \circ s = \tilde{s} \circ q_L^F$ for some $\tilde{s} \in \mathrm{CB}(F/L, F_0/L_0)$ with $\|\tilde{s}\|_{\mathrm{cb}} = \|s\|_{\mathrm{cb}}$. Therefore, by the ideal property for $(\mathfrak{A}, \mathbf{A})$

$$\mathbf{A}_n\big((q_{L_0}^{F_0})_n \circ s_n \circ T \circ r \circ i_{M_0}^{E_0}\big) = \mathbf{A}_n\big(\tilde{s}_n \circ (q_L^F)_n \circ T \circ i_M^E \circ r \circ i_{M_0}^{F_0}\big)$$

$$\leq \|\tilde{s}\|_{\mathrm{cb}} \mathbf{A}_n\big((q_L^F)_n \circ T \circ i_M^E\big) \|r \circ i_{M_0}^{E_0}\|_{\mathrm{cb}} \leq \|s\|_{\mathrm{cb}} \mathbf{A}_n^{\max}(T) \|r\|_{\mathrm{cb}},$$

which yields the ideal property for $(\mathfrak{A}^{\max}, \mathbf{A}^{\max})$.

Let us now check that $\mathfrak{A}^{\max}(E,F)$ is complete when $(\mathfrak{A}, \mathbf{A})$ is a mapping ideal. For that, let $(T_k)_{k=1}^\infty$ be a sequence in $\mathfrak{A}^{\max}(E,F)$ with $\sum_{k=1}^\infty \mathbf{A}_1^{\max}(T_k) < \infty$. Since $\mathbf{A}_1^{\max}$ dominates the cb-norm, the series $\sum_{k=1}^\infty T_k$ converges in $\mathrm{CB}(E,F)$ to an operator T. Let us now show that the convergence also holds with respect to $\mathbf{A}_1^{\max}$. Fix $M \in \mathrm{OFIN}(E)$ and $L \in \mathrm{OCOFIN}(F)$. For $m \in \mathbb{N}$, note that

$$\mathbf{A}_1\Big(q_L^F \circ \Big(\sum_{k=m}^\infty T_k\Big) \circ i_M^E\Big) \leq \sum_{k=m}^\infty \mathbf{A}_1\big(q_L^F \circ T_k \circ i_M^E\big) \leq \sum_{k=m}^\infty \mathbf{A}_1^{\max}(T_k),$$

from where $\mathbf{A}_1^{\max}\big(\sum_{k=m}^\infty T_k\big) \leq \sum_{k=m}^\infty \mathbf{A}_1^{\max}(T_k)$ and therefore $\sum_{k=1}^\infty T_k$ converges to T with respect to $\mathbf{A}_1^{\max}$. $\qquad\square$

Example 8.1.3 The following mapping ideals are maximal.

(i) Completely bounded mappings CB.

(ii) Completely integral mappings $\mathcal{I}$: See Example 8.2.5, below.

(iii) Completely p-summing mappings Π_p: Let $E, F \in \mathrm{OBAN}$, $T \in M_n(\Pi_p(E,F))$ and $\varepsilon > 0$. Assume $(\pi_p)_n(T) = 1$. Since $\Pi_p(E,F)$ is canonically completely isometrically embedded into $\mathrm{CB}(S_p \otimes_{\min} E, S_p[F])$, by a density argument there exist $k \in \mathbb{N}$ and $x \in M_n(S_p^k \otimes_{\min} E)$ with $\|x\| = 1$ such that

$$\left\| (Id_{S_p^k} \otimes T)(x) \right\|_{M_n(S_p^k[F])} > 1 - \varepsilon,$$

Since $S_p^k[F]$ is completely isometrically embedded into $S_p^k[F''] = (S_{p'}[F'])'$, there exists $y \in M_\ell(S_{p'}^k[F'])$ with $\|y\| = 1$ and such that

$$\left\| \langle\langle y, (id_{S_p^k} \otimes T)(x) \rangle\rangle \right\|_{M_{n\ell}} > 1 - \varepsilon.$$

Consider y as an $k\ell \times k\ell$ matrix over F', and let L be the intersection of these $(k\ell)^2$ functionals in F'. Note that $L \in \mathrm{OCOFIN}(F)$, and that y is an $k\ell \times k\ell$ matrix over the annihilator L^0 of L in F. Since $(F/L) = L^0$ completely isometrically, it follows that $S_{p'}^k[F/L] = S_{p'}^k[L^0]$ completely isometrically as well. Therefore, the norm of y as an element of $M_\ell(S_{p'}^k[(F/L)'])$ is also equal to one. If M is the finite-dimensional subspace of E spanned by the $(nk)^2$ vectors that constitute x, it follows that

$$(\pi_p)_n\big((q_L^F)_n \circ T \circ i_M^E\big) > 1 - \varepsilon.$$

8.2 The Representation Theorem

Definition 8.2.1 A mapping ideal $(\mathfrak{A}, \mathbf{A})$ and a finitely-generated o.s. tensor norm α (on ONORM) are said to be *associated*, denoted $(\mathfrak{A}, \mathbf{A}) \sim \alpha$ if for every $M, N \in$ OFIN we have a complete isometry

$$\mathfrak{A}(M, N) = M' \otimes_\alpha N,$$

given by the canonical map $T \mapsto z_T$.

Remark 8.2.2 Notice that since this definition is based only on finite-dimensional spaces; two different mapping ideals can be associated to the same o.s. tensor norm. In particular, $\mathfrak{A} \sim \alpha$ if and only if $\mathfrak{A}^{\max} \sim \alpha$.

For example, CB, $\mathcal{A} \sim$ min or $\mathcal{N}, \mathcal{I} \sim$ proj. Also, by the mere definition, $\mathcal{N}_p \sim d_p$. Additionally, by Eq. (7.1.2), $\Pi_p \sim (d_{p'})' = g_{p'}^*$.

Observe that the constructions in Definitions 3.1.1 and 8.1.1 establish a one-to-one correspondence between maximal mapping ideals and finitely-generated o.s. tensor norms. The main result of this chapter is the following theorem, which shows that the finite-dimensional duality that defines this correspondence can be extended to the infinite-dimensional setting.

Theorem 8.2.3 (Representation Theorem for Maximal Mapping Ideals) *Let $(\mathfrak{A}, \mathbf{A})$ be a maximal mapping ideal associated to a finitely-generated o.s. tensor norm α. Then for any $E, F \in$ OBAN there is a complete isometry*

$$\mathfrak{A}(E, F) = \big(E \otimes_{\alpha'} F'\big)' \cap \mathrm{CB}(E, F). \tag{8.2.1}$$

Proof In order to prove (8.2.1), we need to show that for $T \in M_n(\mathrm{CB}(E, F))$, the following holds: T belongs to $M_n(\mathfrak{A}(E, F))$ if and only if the associated matrix of bilinear maps $\beta_{\kappa_F \circ T}$ is in $M_n\big((E \otimes_{\alpha'} F')'\big)$, with equal norms. Note that the inequality $\big\|\beta_{\kappa_F \circ T}\big\| \leq C$ is equivalent to

$$\big\|\langle\langle \beta_{\kappa_F \circ T}, z\rangle\rangle\big\|_{M_{nm}} \leq C(\alpha')_m(z; E, F') \text{ for all } z \in M_m(E \otimes_{\alpha'} F'). \tag{8.2.2}$$

By the maximality of $\mathfrak{A}$, it follows that $T \in M_n(\mathfrak{A}(E, F))$ with $\mathbf{A}_n(T) \leq C$ if and only if

$$\mathbf{A}_n\big((q_L^F)_n \circ T \circ i_M^E\big) \leq C \qquad \text{for any } M \in \mathrm{OFIN}(E) \text{ and } L \in \mathrm{OCOFIN}(F). \tag{8.2.3}$$

Since $\mathfrak{A}(M, F/L) = (M \otimes_{\alpha'} L^0)'$ completely isometrically, and recalling that L^0 varies over all spaces in $\mathrm{OFIN}(F')$ when L varies over all spaces in $\mathrm{OCOFIN}(F)$, condition (8.2.3) is equivalent to

$$\Big\|\langle\langle \beta_{(q_L^F)_n \circ T \circ i_M^E}, z\rangle\rangle\Big\|_{M_{mn}} \leq C(\alpha')_m(z; M, L^0)$$

$$\text{for any } M \in \mathrm{OFIN}(E), \ L^0 \in \mathrm{OFIN}(F') \text{ and } z \in M_m(M \otimes_{\alpha'} L^0). \tag{8.2.4}$$

Notice that for $z \in M_m(M \otimes_{\alpha'} L^0)$,

$$\langle\langle \beta_{\kappa_F \circ T}, z\rangle\rangle = \langle\langle \beta_{(q_L^F)_n \circ T \circ i_M^E}, z\rangle\rangle.$$

Therefore (8.2.4) is equivalent to (8.2.2) because α' is finitely-generated, finishing the proof. $\square$

In fact the converse is also true. We write the statement in the following proposition and leave the proof as an exercise. Recall that in Example 7.1.4 we define a mapping ideal "dual" to a given o.s. tensor norm.

Proposition 8.2.4 *Let α be a finitely-generated o.s. tensor norm. The mapping ideal* $(\mathfrak{A}_{\alpha'}, \mathbf{A}_{\alpha'})$ *(given by*

$$\mathfrak{A}_{\alpha'}(E, F) := \big(E \otimes_{\alpha'} F'\big)' \cap \mathrm{CB}(E, F) \tag{8.2.5}$$

for any $E, F \in$ OBAN) is maximal and its associated o.s. tensor norm is α.

We use this fact to see, in the following example, that the ideal of completely integral mappings is maximal.

Example 8.2.5 $\mathcal{I}$ is a maximal mapping ideal.

Proof We will give two proofs of this (for the second, we present only a sketch of it). Let's start with the first one.

Recall that the o.s. tensor norm associated to $\mathcal{I}$ is proj. By the previous proposition, to see that $\mathcal{I}$ is maximal we need to check that for every $E, F \in$ OBAN there is a complete isometry

$$\mathcal{I}(E, F) = \left(E \otimes_{\min} F'\right)' \cap \mathrm{CB}(E, F). \qquad (8.2.6)$$

By [34, Lem. 12.3.3] the canonical inclusion

$$S_0 : \mathcal{I}(E, F) \hookrightarrow (E \otimes_{\min} F')'$$

is a complete isometry. Thus, it remains to see that if $\Psi \in (E \otimes_{\min} F')' \cap \mathrm{CB}(E, F)$ then $\Psi = S_0(T)$, for some $T \in \mathcal{I}(E, F)$. Without loss of generality, we suppose that $\|\Psi\|_{(E \otimes_{\min} F')'} < 1$. By Remark 5.2.2 the following inclusion is completely isometric:

$$E \otimes_{\min} F' \hookrightarrow (E' \otimes_{\mathrm{proj}} F)' = \mathrm{CB}(E', F').$$

Then, by the Averson-Wittstock Hahn-Banach theorem [34, Thm. 4.1.5], Ψ extends isometrically to $(E' \otimes_{\mathrm{proj}} F)''$. Therefore, Goldstine's theorem provides us of a net $(\Psi_\gamma)_\gamma$ such that $\|\Psi_\gamma\|_{E' \otimes_{\mathrm{proj}} F} < 1$, which w^*-converges to Ψ (that is in the topology $\sigma((E' \otimes_{\mathrm{proj}} F)'', (E' \otimes_{\mathrm{proj}} F)')$). We consider $\Phi(\Psi_\gamma)_\gamma \subset \mathcal{N}(E, F)$ with $\nu(\Phi(\Psi_\gamma)) < 1$, where Φ is the mapping defined in (7.1.1). Note that given $x \in E$, $y' \in F'$

$$\langle y', \Phi(\Psi_\gamma) \rangle \to \langle \Psi, x \otimes y' \rangle.$$

Now, the expression $\langle T(x), y' \rangle := \langle \Psi, x \otimes y' \rangle$ defines a mapping $T \in \mathrm{CB}(E, F)$ satisfying that $\Phi(\Psi_\gamma)$ converges to T in the point-weak topology. Finally, appealing to [34, Lem. 12.3.1] we get that T is completely integral and $\Psi = S_0(T)$. This concludes the first proof.

We now present a sketch of an alternative argument for the fact that $\mathcal{I}$ is maximal, whose perspective could be enriching. First, observe that [23, Ex. 17.1] also holds in the operator space setting. That is: suppose that for each $\gamma \in \Gamma$, $(\mathfrak{A}^\gamma, \mathbf{A}^\gamma)$ is a maximal mapping ideal. Define for $E, F \in$ OBAN and $T \in M_n(\mathrm{CB}(E, F))$,

$$\mathbf{A}_n(T) = \sup\left\{\mathbf{A}_n^\gamma(T) : \gamma \in \Gamma\right\} \quad \text{and}$$

$$\mathfrak{A}(E, F) = \{S \in \mathrm{CB}(E, F) : \mathbf{A}_1(S) < \infty\}.$$

It is easy to see that $(\mathfrak{A}, \mathbf{A})$ is also a maximal mapping ideal. Now observe that [34, Lem. 12.3.1] says that the ideal of completely integral mappings can be obtained by applying the above procedure to the ideals $(\mathfrak{A}^G, \mathbf{A}^G)$ defined by the completely isometric embeddings

$$\mathfrak{A}^G(E, F) \hookrightarrow \mathrm{CB}(E \otimes_{\min} G, F \otimes_{\mathrm{proj}} G),$$

where G ranges over all Banach operator spaces (or just the finite-dimensional ones). Since each $\mathfrak{A}^G$ is clearly maximal, it follows that so is $\mathcal{L}$. This concludes the sketch of the second proof. $\qquad\square$

Proposition 8.2.6 *Let* $(\mathfrak{A}, \mathbf{A})$ *and* $(\mathfrak{B}, \mathbf{B})$ *be Banach mapping ideals. Suppose that for every* $M, N \in \mathrm{OFIN}$ *we have that the inclusion* $\|\mathfrak{A}(M, N) \hookrightarrow \mathfrak{B}(M, N)\|_{\mathrm{cb}} \leq c$. *Then* $\mathfrak{A}^{\max} \subset \mathfrak{B}^{\max}$ *and* $\|\mathfrak{A}^{\max}(E, F) \hookrightarrow \mathfrak{B}^{\max}(E, F)\|_{\mathrm{cb}} \leq c$ *for every* $E, F \in$ OBAN.

Proof Let $T \in M_n(\mathrm{CB}(E, F))$. Fix $M \in \mathrm{OFIN}(E)$ and $L \in \mathrm{OCOFIN}(F)$. From the assumption, note that $\mathbf{B}_n((q_L^F)_n \circ T \circ i_M^E) \leq c\mathbf{A}_n((q_L^F)_n \circ T \circ i_M^E)$. Taking the supremum over all M and L yields $\mathbf{B}_n^{\max}(T) \leq c\mathbf{A}_n^{\max}(T)$, giving the result. $\qquad\square$

Proposition 8.2.7 *Let* $(\mathfrak{A}, \mathbf{A})$ *be a mapping ideal associated to a finitely-generated o.s. tensor norm* α. *Then* $(\mathfrak{A}^{\max}, \mathbf{A}^{\max})$ *is the largest mapping ideal associated to* α.

Proof The ideal property of $(\mathfrak{A}, \mathbf{A})$ yields that the identity $\mathfrak{A}(E, F) \to \mathfrak{A}^{\max}(E, F)$ is completely contractive for any $E, F \in$ OBAN. For $M, N \in$ OFIN we additionally get that the identity $\mathfrak{A}^{\max}(M, N) \to \mathfrak{A}(M, N)$ is also completely contractive since for $T \in \mathrm{CB}(M, N)$ we have that $T = (q_{\{0\}}^N)_n \circ T \circ i_M^M$. Therefore, the identity $\mathfrak{A}^{\max}(M, N) \to \mathfrak{A}(M, N)$ is a complete isomorphism and thus $(\mathfrak{A}^{\max}, \mathbf{A}^{\max})$ is associated to α.

Let $(\mathfrak{B}, \mathbf{B})$ be any mapping ideal associated to α, and let $T \in M_n(\mathfrak{B}(E, F))$. Let $M \in \mathrm{OFIN}(E)$ and $L \in \mathrm{OCOFIN}(F)$. By the ideal property for $(\mathfrak{B}, \mathbf{B})$ we have that

$$\mathbf{A}_n((q_L^F)_n \circ T \circ i_M^E) = \mathbf{B}_n((q_L^F)_n \circ T \circ i_M^E) \leq \mathbf{B}_n(T),$$

from where $\mathbf{A}_n^{\max}(T) \leq \mathbf{B}_n(T)$, showing that $\mathfrak{B}(E, F) \subseteq \mathfrak{A}^{\max}(E, F)$ (and in fact the inclusion $\mathfrak{B}(E, F) \to \mathfrak{A}^{\max}(E, F)$ is completely contractive). $\qquad\square$

Remark 8.2.8 Given a finite dimensional operator space M and an arbitrary $E \in$ OBAN there is a canonical complete isometry:

$$M \otimes_{\mathrm{proj}} E \hookrightarrow (M' \otimes_{\mathrm{min}} E')'.$$

Indeed, since M is finite dimensional we have $M \otimes_{\mathrm{proj}} E = \mathcal{N}(M', E) = \mathcal{I}(M', E)$. Now, the result follows through the embedding S_0 referred to in Example 8.2.5. By transposition, there is also the following complete isometry:

$$E \otimes_{\mathrm{proj}} M \hookrightarrow (E' \otimes_{\mathrm{min}} M')'.$$

Another classical version of the Representation Theorem states, in particular, that a maximal ideal taking values in a dual space is itself a dual space. To get the operator space version, we need the additional hypothesis of local reflexivity.

Corollary 8.2.9 *Let* $(\mathfrak{A}, \mathbf{A})$ *be a maximal mapping ideal and* α *its associated finitely-generated o.s. tensor norm. If* F *is locally reflexive, then*

$$\mathfrak{A}(E, F') = (E \otimes_{\alpha'} F)'$$

Proof Consider the diagram

$$
\begin{array}{ccc}
\varphi \in (E \otimes_{\alpha'} F)' & \longrightarrow & (E \otimes_{\mathrm{proj}} F)' =\!=\!=\!= \mathrm{CB}(E, F') \\
\uparrow & & \uparrow \\
\downarrow & & \downarrow \\
\varphi^{\wedge} \in (E \otimes_{\alpha'} F'')' & \longrightarrow & (E \otimes_{\mathrm{proj}} F'')'.
\end{array}
$$

The vertical arrows are complete isometries thanks to the Extension Lemma 4.2.1, whereas the horizontal arrows are completely contractive because $\alpha' \leq \mathrm{proj}$. The desired conclusion follows from (8.2.5). $\square$

Note that the hypothesis of local reflexivity cannot be omitted from the previous result. Indeed, taking $\mathfrak{A} = \mathcal{I}$ (the mapping ideal of completely integral mappings) which is associated to the o.s. tensor norm proj, we know by [34, Thm. 14.3.1] that the equality $\mathcal{I}(E, F') = (E \otimes_{\mathrm{min}} F)'$ holds for every $E \in \mathrm{OBAN}$ if and only if F is locally reflexive.

On the other hand, whenever α' is an $\mathcal{E}(\lambda)$-o.s. tensor norm the Extension Lemma 4.2.1 is valid without the hypothesis of local reflexivity. Thus, the previous corollary also holds without this hypothesis.

Along similar lines, in this same setting we can get a version of [23, Thm. 17.15].

Theorem 8.2.10 *Let* $(\mathfrak{A}, \mathbf{A})$ *be a maximal mapping ideal associated to a finitely-generated o.s. tensor norm* α. *Suppose that* α' *is an* $\mathcal{E}(\lambda)$-*o.s. tensor norm. Then for any* $E, F \in \mathrm{OBAN}$ *and* $T \in \mathrm{CB}(E, F)$, *the following are equivalent:*

(a) $T \in \mathfrak{A}(E, F)$.
(b) For any $G \in \mathrm{OBAN}$ *(or only* $G = F'$*) the map*

$$T \otimes id_G : E \otimes_{\alpha'} G \to F \otimes_{\mathrm{proj}} G$$

is bounded.

In this case,

$$\mathbf{A}(T) = \left\| T \otimes id_{F'} : E \otimes_{\alpha'} F' \to F \otimes_{\mathrm{proj}} F' \right\|$$

$$\geq \left\| T \otimes id_G : E \otimes_{\alpha'} G \to F \otimes_{\mathrm{proj}} G \right\|.$$

Proof Suppose that $T \in \mathfrak{A}(E, F)$. Let $\varphi \in (F \otimes_{\mathrm{proj}} G)' = \mathrm{CB}(F, G')$. Note that $L_{\varphi} \circ T \in \mathfrak{A}(E, G') = (E \otimes_{\alpha'} G)'$, where we are using that the previous corollary holds in general under the present conditions on α'. We then have for $z \in E \otimes G$, using [23, Eqn. 17.15.(1)],

$$|\langle \varphi, (T \otimes id_G)(z)\rangle| = |\langle \beta_{L_\varphi \circ T}, z\rangle| \leq \mathbf{A}(L_\varphi \circ T)\alpha'(z; E, G)$$
$$\leq \|\varphi\| \mathbf{A}(T)\alpha'(z; E, G).$$

Taking the supremum over $\|\varphi\| \leq 1$ yields

$$\mathrm{proj}((T \otimes id_G)(z); F, G) \leq \mathbf{A}(T)\alpha'(z; E, G).$$

Now, assume that (b) is satisfied for $G = F'$. Recall that from the Representation Theorem, $\mathfrak{A}(E, F)$ coincides with $(E \otimes_{\alpha'} F')' \cap \mathrm{CB}(E, F)$. Since $T \in \mathrm{CB}(E, F)$, it will suffice to show that $\beta_{\kappa_F \circ T} \in (E \otimes_{\alpha'} F')'$. If we now let $z \in E \otimes F'$, using [23, Eqn. 17.15.(2)],

$$|\langle \beta_{\kappa_F \circ T}, z\rangle| = |\langle \mathrm{tr}_F, (T \otimes id_{F'})(z)\rangle| \leq \mathrm{proj}((T \otimes id_{F'})(z); F, F')$$
$$\leq \alpha'(z; E, F') \left\| T \otimes id_{F'} : E \otimes_{\alpha'} F' \to F \otimes_{\mathrm{proj}} F' \right\|.$$

$$\square$$

Once we have the theorem above, we get a version of [23, Prop. 17.20]. Notice that this is a generalization of [1, Prop. 4], where the same result is obtained for some λ-o.s. tensor norms.

Proposition 8.2.11 *If α is an $\mathcal{E}(\lambda)$-o.s. tensor norm and $E, F \in$ OBAN, one of which has the OAP, then the natural map*

$$i : E \widehat{\otimes}_\alpha F \to E \widehat{\otimes}_{\min} F$$

is injective.

Proof Suppose that F has the OAP. Let $z \in E \widehat{\otimes}_\alpha F$ be such that $i(z) = 0$. Observe that we need to show that $\langle \varphi, z\rangle = 0$ whenever $\varphi \in (E \widehat{\otimes}_\alpha F)' \hookrightarrow \mathrm{CB}(E, F')$. Since α is finitely-generated (see Definition 4.2.3), we have $\alpha = (\alpha')'$ and α' is associated to the maximal mapping ideal $\mathfrak{A}_\alpha$. Also, by the comments after Corollary 8.2.9, $\mathfrak{A}_\alpha(E, F') = (E \otimes_\alpha F)'$ because α is an $\mathcal{E}(\lambda)$-o.s. tensor norm. Therefore the previous theorem applies, so

$$L_\varphi \otimes id_F : E \widehat{\otimes}_\alpha F \to F' \widehat{\otimes}_{\mathrm{proj}} F$$

is continuous. In the diagram

$$
\begin{array}{ccc}
E \widehat{\otimes}_\alpha F & \xrightarrow{\ i\ } & E \widehat{\otimes}_{\min} F \\
{\scriptstyle L_\varphi \widehat{\otimes}_{\mathrm{proj}} id_F} \downarrow & & \downarrow {\scriptstyle L_\varphi \widehat{\otimes}_{\min} id_F} \\
F' \widehat{\otimes}_{\mathrm{proj}} F & \longrightarrow & F' \widehat{\otimes}_{\min} F
\end{array}
$$

the bottom row is injective because F has OAP [34, Thm. 11.2.5], which implies

$$(L_\varphi \widehat{\otimes}_{\alpha,\mathrm{proj}} id_F)(z) = 0 \in F' \widehat{\otimes}_{\mathrm{proj}} F.$$

Once again using [23, Eqn. 17.15.(2)], we get

$$\langle \varphi, z \rangle = \langle \mathrm{tr}_F, (L_\varphi \widehat{\otimes}_{\alpha,\mathrm{proj}} id_F)(z) \rangle = 0.$$

$\square$

Example 8.2.12 For the Haagerup o.s. tensor norm h consider as in Example 7.1.4 the mapping ideal $\mathfrak{A}_h$. Then, $\mathfrak{A}_h$ is a maximal mapping ideal defined as

$$\mathfrak{A}_h(E, F) := (E \otimes_h F')' \cap \mathrm{CB}(E, F)$$

for every $E, F \in \mathrm{OBAN}$.

Since $\alpha' = h$ is a λ-o.s. tensor norm, we have

$$\mathfrak{A}_h(E, F') = (E \otimes_h F)'$$

for every $E, F \in \mathrm{OBAN}$. Also, recall that the dual of $E \otimes_h F$ is identified with the space $\mathrm{MB}(E \times F)$ of multiplicatively bounded bilinear mappings (see, for instance, [34, Prop. 9.2.2]) then we have the following complete isometry:

$$\mathfrak{A}_h(E, F') = \mathrm{MB}(E \times F).$$

Additionally, there is a completely isometric canonical identification $(E \otimes_h F)' = \Gamma_r(E, F')$ [32, Eqn. (5.8)], which shows that $\mathfrak{A}_h(E, F') = \Gamma_r(E, F')$. In the general case,

$$\mathfrak{A}_h(E, F) := (E \otimes_h F')' \cap \mathrm{CB}(E, F) = \Gamma_r(E, F'') \cap \mathrm{CB}(E, F)$$

which means that $T : E \to F$ belongs to $\mathfrak{A}_h(E, F)$ if and only if $\kappa_F T \in \Gamma_r(E, F'')$. There are analogous results for the o.s. tensor norm h^t. The mapping ideal $\mathfrak{A}_{h^t}$ is also maximal, we have $\mathfrak{A}_{h^t}(E, F') = (E \otimes_{h^t} F)'$, $\mathfrak{A}_{h^t}(E, F') = \mathrm{MB}(F \times E)$, and $\mathfrak{A}_{h^t}(E, F') = \Gamma_c(E, F')$ (see [32, Thm. 5.3]), so in the general case $T : E \to F$ belongs to $\mathfrak{A}_{h^t}(E, F)$ if and only if $\kappa_F T \in \Gamma_c(E, F'')$.

From the previous example and Theorem 8.2.10, we get the following:

Corollary 8.2.13 *For any $E, F \in \mathrm{OBAN}$ and $T \in \mathrm{CB}(E, F)$, the following are equivalent:*

(a) $\kappa_F T \in \Gamma_r(E, F'')$ (resp. $\kappa_F T \in \Gamma_c(E, F'')$).
(b) For any $G \in \mathrm{OBAN}$ (or only $G = F'$), the map $T \otimes id_G : E \otimes_h G \to F \otimes_{\mathrm{proj}} G$ (resp. $T \otimes id_G : E \otimes_{h^t} G \to F \otimes_{\mathrm{proj}} G$) is bounded.

Moreover, in this case we have $\gamma_2^r(\kappa_F T) = \big\| T \otimes id_{F'} : E \otimes_h F' \to F \otimes_{\mathrm{proj}} F' \big\|$
(resp. $\gamma_2^c(\kappa_F T) = \big\| T \otimes id_{F'} : E \otimes_{h^t} F' \to F \otimes_{\mathrm{proj}} F' \big\|$*).*

Furthermore, similar representations are valid for $h + h^t$ and $h \cap h^t$.

Example 8.2.14 Recall that $\mathfrak{A}_{h+h^t}$ is a maximal mapping ideal defined as

$$\mathfrak{A}_{h+h^t}(E, F) := (E \otimes_{h+h^t} F')' \cap \mathrm{CB}(E, F)$$

for every $E, F \in \mathrm{OBAN}$. Since $h+h^t$ is an $\mathcal{E}(\lambda)$-o.s. tensor norm, by Corollary 8.2.9 we have

$$\mathfrak{A}_{h+h^t}(E, F') = (E \otimes_{h+h^t} F)'$$

for every $E, F \in \mathrm{OBAN}$. Furthermore, dualizing the canonical quotient

$$(E \otimes_h F) \oplus_1 (E \otimes_{h^t} F) \twoheadrightarrow E \otimes_{h+h^t} F$$

yields a completely isometric inclusion

$$(E \otimes_{h+h^t} F)' \hookrightarrow \Delta\big((E \otimes_h F)' \oplus_\infty (E \otimes_{h^t} F)'\big),$$

where Δ denotes the diagonal, which together with the identifications above produces a completely isometric inclusion

$$\mathfrak{A}_{h+h^t}(E, F') \hookrightarrow \Delta\big(\Gamma_r(E, F') \oplus_\infty \Gamma_c(E, F')\big).$$

In the general case, the same arguments yield a completely isometric inclusion

$$\mathfrak{A}_{h+h^t}(E, F) \hookrightarrow \Delta\big(\Gamma_r(E, F'') \oplus_\infty \Gamma_c(E, F'')\big) \cap \mathrm{CB}(E, F).$$

Let us verify that this is in fact an equality:

$$\mathfrak{A}_{h+h^t}(E, F) = \Delta\big(\Gamma_r(E, F'') \oplus_\infty \Gamma_c(E, F'')\big) \cap \mathrm{CB}(E, F).$$

Indeed, if $T \in \mathrm{CB}(E, F)$ satisfies $\kappa_F T \in \Gamma_r(E, F'')$ and $\kappa_F T \in \Gamma_c(E, F'')$, by Example 8.2.12 we have $T \in (E \otimes_h F')' \cap \mathrm{CB}(E, F)$ and $T \in (E \otimes_{h^t} F')' \cap \mathrm{CB}(E, F)$. If we denote by $L_T : E \otimes F' \to \mathbb{C}$ the linear map induced by T, we then have that $L_T : E \otimes_h F' \to \mathbb{C}$ and $L_T : E \otimes_{h^t} F' \to \mathbb{C}$ are bounded, hence so is the map

$$L_T \oplus L_T : (E \otimes_h F') \oplus_1 (E \otimes_{h^t} F') \to \mathbb{C},$$

and therefore $L_T : (E \otimes_{h+h^t} F') \to \mathbb{C}$ is bounded, which means $T \in (E \otimes_{h+h^t} F')' \cap \mathrm{CB}(E, F) = \mathfrak{A}_{h+h^t}(E, F)$.

Analogously, $\mathfrak{A}_{h\cap h^t}$ is a maximal mapping ideal defined as

$$\mathfrak{A}_{h\cap h^t}(E, F) := (E \otimes_{h\cap h^t} F')' \cap \mathrm{CB}(E, F)$$

for every $E, F \in \mathrm{OBAN}$. Since $h \cap h^t$ is an $\mathcal{E}(\lambda)$-o.s. tensor norm, we have

$$\mathfrak{A}_{h\cap h^t}(E, F') = (E \otimes_{h\cap h^t} F)'$$

for every $E, F \in \mathrm{OBAN}$. Moreover, dualizing the complete injection

$$E \otimes_{h\cap h^t} F \hookrightarrow (E \otimes_h F) \oplus_\infty (E \otimes_{h^t} F)$$

yields the complete quotient

$$(E \otimes_h F)' \oplus_1 (E \otimes_{h^t} F)' \twoheadrightarrow (E \otimes_{h\cap h^t} F)'$$

which, together with the identifications above, translates to

$$\Gamma_r(E, F') \oplus_1 \Gamma_c(E, F') \twoheadrightarrow \mathfrak{A}_{h\cap h^t}(E, F').$$

8.2.1　Consequences of the Representation Theorem

Combining the Representation Theorem 8.2.3 and Corollary 8.2.9 with the Duality Theorem 5.2.1 we obtain:

Theorem 8.2.15 (Embedding Theorem)　*Let* $(\mathfrak{A}, \mathbf{A})$ *be a maximal mapping ideal associated to the o.s. tensor norm* α. *Then for any* $E, F \in \mathrm{OBAN}$ *there are complete isometries:*

$$E \otimes_{\overleftarrow{\alpha}} F \hookrightarrow \mathfrak{A}(E', F)$$

$$E' \otimes_{\overleftarrow{\alpha}} F \hookrightarrow \mathfrak{A}(E, F) \qquad \text{whenever } E \text{ is locally reflexive}$$

$$E' \otimes_{\overleftarrow{\alpha}} F' \hookrightarrow \mathfrak{A}(E, F') \qquad \text{whenever } E \text{ and } F \text{ are locally reflexive.}$$

Remark 8.2.16　As in the comments after the Duality Theorem 5.2.1 and Corollary 8.2.9, if α' is an $\mathcal{E}(\lambda)$-o.s. tensor norm then the Embedding Theorem 8.2.15 is valid without the hypothesis of local reflexivity. In particular, this holds in the cases $\overleftarrow{\alpha} = \overleftarrow{\min} = \min$ with $\mathfrak{A} = \mathrm{CB}$ and $\overleftarrow{\alpha} = \overleftarrow{h} = h$ with $\mathfrak{A} = \mathfrak{A}_h$.

In the classical setting the dual of a maximal ideal is also maximal, and their associated tensor norms are transposes of each other. In the operator space setting, in order to get a similar result we once again need the additional hypothesis of local reflexivity.

Corollary 8.2.17 *Let $(\mathfrak{A}, \mathbf{A})$ be a maximal mapping ideal associated to the finitely-generated o.s. tensor norm α. Then $\mathfrak{A}^{dual}$ is associated to α^t. Moreover, if $(\mathfrak{B}, \mathbf{B})$ is the maximal mapping ideal associated to α^t and E is a locally reflexive Banach operator space, then $\mathfrak{B}(E, F) = \mathfrak{A}^{\mathrm{dual}}(E, F)$.*

Proof If $\mathfrak{A} \sim \alpha$ and M, N are finite dimensional operator spaces then $\mathfrak{A}^{\mathrm{dual}}(M, N) = M' \otimes_{\alpha^t} N$. This shows that $\mathfrak{A}^{\mathrm{dual}} \sim \alpha^t$. Also, by the Representation Theorem 8.2.3 and Corollary 8.2.9 we have complete isometries

$$\mathfrak{B}(E, F) = (E \otimes_{(\alpha^t)'} F')' \cap \mathrm{CB}(E, F) = \left\{ T \in \mathrm{CB}(E, F) \, : \, \beta_T \in (E \otimes_{(\alpha^t)'} F')' \right\}$$

$$= \left\{ T \in \mathrm{CB}(E, F) \, : \, \beta_{T'} \in (F' \otimes_{\alpha'} E)' \right\}$$

$$= \left\{ T \in \mathrm{CB}(E, F) \, : \, T' \in \mathfrak{A}(F', E') \right\} = \mathfrak{A}^{\mathrm{dual}}(E, F)$$

and therefore, for $S \in M_n(\mathfrak{B}(E, F))$ we have $\mathbf{B}_n(S) = \mathbf{A}_n(S')$. This is the desired conclusion. $\square$

If $(\mathfrak{A}, \mathbf{A})$ is a maximal mapping ideal, it should be noted that $\mathfrak{A}^{\mathrm{dual}}$ is not necessarily maximal as it happens in the Banach space case. Indeed, $\mathfrak{A} = \mathcal{I}$ the maximal mapping ideal of completely integral mappings, which is associated to the o.s. tensor norm proj, gives us the example. To see this, note that $\mathcal{I}^{\mathrm{dual}}(M, N) = M' \otimes_{\mathrm{proj}} N$, for every $M, N \in \mathrm{OFIN}$. Thus, if $\mathcal{I}^{\mathrm{dual}}$ were a maximal mapping ideal it should coincide with $\mathcal{I}$ (since $\mathrm{proj}^t = \mathrm{proj}$). If W is any non-locally reflexive o.s. then by [34, Thm. 14.3.1] there is $L \in \mathrm{OFIN}$ such that the inclusion $\mathcal{N}(W, L') \hookrightarrow \mathcal{I}(W, L')$ is not isometric. Let $T \in \mathcal{N}(W, L')$ such that $\iota(T) < \nu(T)$. We have that $T' \in \mathcal{N}(L, W') = \mathcal{I}(L, W')$ (since L is finite-dimensional) and

$$\iota^{\mathrm{dual}}(T) = \iota(T') = \nu(T') = \nu(T) > \iota(T).$$

The situation of the previous example changes when α' is an $\mathcal{E}(\lambda)$-o.s. tensor norm. Indeed, in this case Corollary 8.2.9 is valid without requiring a local reflexivity hypothesis and hence we have the following consequence of Corollary 8.2.17:

Remark 8.2.18 If $(\mathfrak{A}, \mathbf{A})$ is a maximal mapping ideal associated to the finitely-generated o.s. tensor norm α and α' is an $\mathcal{E}(\lambda)$-o.s. tensor norm, then $\mathfrak{A}^{\mathrm{dual}}$ is a maximal mapping ideal associated with α^t.

Recall that for finite-dimensional spaces E and F and linear maps $T : M \to N$, $S : N \to M$, their trace duality pairing is given by

$$\langle T, S \rangle = \mathrm{tr}(S \circ T).$$

More generally, for $T \in M_n(\mathrm{CB}(M, N))$ and $S \in M_m(\mathrm{CB}(N, M))$, their trace duality (matrix) pairing is the matrix in M_{nm} given by

$$\langle\langle T, S \rangle\rangle = [\mathrm{tr}(S_{kl} \circ T_{ij})].$$

Definition 8.2.19 Let $(\mathfrak{A}, \mathbf{A})$ be a mapping ideal. For $M, N \in$ OFIN, $T \in M_n(\mathrm{CB}(M, N))$ and $S \in M_m(\mathrm{CB}(N, M))$, we denote by $S \circ T$ the induced $mn \times mn$ matrix of mappings $M \to M$. For $T \in M_n(\mathrm{CB}(E, F))$, define

$$\mathbf{A}_n^*(T) = \sup \left\{ \left\| \langle\langle (q_L^F)_n \circ T \circ i_M^E, S\rangle\rangle \right\|_{M_{mn}} : \right.$$
$$\left. M \in \mathrm{OFIN}(E), L \in \mathrm{OCOFIN}(F), S \in M_m(\mathrm{CB}(F/L, M)), \mathbf{A}_m(S) \leq 1 \right\}$$

and

$$\mathfrak{A}^*(E, F) = \left\{ T \in \mathrm{CB}(E, F) : \mathbf{A}_1^*(T) < \infty \right\}.$$

As the following proposition shows, $(\mathfrak{A}^*, \mathbf{A}^*)$ is a maximal mapping ideal which is associated with the so-called adjoint o.s. tensor norm. Thus, we call $(\mathfrak{A}^*, \mathbf{A}^*)$ *the adjoint mapping ideal* of $(\mathfrak{A}, \mathbf{A})$.

Proposition 8.2.20 *Let $(\mathfrak{A}, \mathbf{A})$ be a mapping ideal associated to the finitely-generated o.s. tensor norm α. Then $(\mathfrak{A}^*, \mathbf{A}^*)$ is a maximal mapping ideal, and it is associated with the adjoint o.s. tensor norm $\alpha^* := (\alpha^t)'$.*

Proof Let $(\mathfrak{B}, \mathbf{B})$ be the maximal mapping ideal associated to α^*. Now we show that $(\mathfrak{B}, \mathbf{B})$ is in fact $(\mathfrak{A}^*, \mathbf{A}^*)$. First, note that, for $M, N \in$ OFIN,

$$\left(\mathfrak{A}(N, M)\right)' = (N' \otimes_\alpha M)' = (M \otimes_{\alpha^t} N') = M' \otimes_{(\alpha^t)'} N = M' \otimes_{\alpha^*} N.$$

Hence,

$$\mathfrak{B}(M, N) = \left(\mathfrak{A}(N, M)\right)'$$

completely isometrically, where the duality pairing is given by the canonical trace duality. This means that for $T \in M_n(\mathrm{CB}(M, N))$ we have

$$\mathbf{B}_n(T) = \sup\{\|\langle\langle T, S\rangle\rangle\|_{M_{nm}} : S \in M_m(\mathfrak{A}(N, M)), \mathbf{A}_m(S) \leq 1\}.$$

Since $\mathfrak{B}$ is maximal, it follows that for any $E, F \in$ OBAN, and $T \in M_n(\mathrm{CB}(E, F))$,

$$\mathbf{B}_n(T) = \sup \left\{ \left\| \langle\langle (q_L^F)_n \circ T \circ i_M^E, S\rangle\rangle \right\|_{M_{mn}} : \right.$$
$$\left. M \in \mathrm{OFIN}(E), L \in \mathrm{OCOFIN}(F), S \in M_m(\mathfrak{A}(F/L, M)), \mathbf{A}_m(S) \leq 1 \right\}$$

which coincides with the definition of $\mathbf{A}_n^*(T)$. $\qquad\square$

As an example of the previous proposition we have $\Pi_p = [(\mathcal{N}_{p'})^{\mathrm{dual}}]^*$. Recall that Π_p is maximal (see Example 8.1.3) and therefore this follows by the fact that $\Pi_p \sim g_{p'}^*$ and $\mathcal{N}_{p'} \sim d_{p'}$, which implies $\mathcal{N}_{p'}^{\mathrm{dual}} \sim (d_{p'})^t = g_{p'}$.

Chapter 9
Minimal Operator Space Mapping Ideals

In the theory Banach spaces another important topic regarding the interplay between tensor products and ideals is the class of minimal operator ideals [23, Sec. 22]. In this chapter we construct an operator space analogue of this concept; an instance of this procedure already appears in the definition of nuclear mappings, as will become clear in Theorem 9.2.3 below.

9.1 The Minimal Hull

Given a Banach mapping ideal $\mathfrak{A}$, we will describe the smallest Banach mapping ideal which coincides with $\mathfrak{A}$ on OFIN. This mapping ideal is called the *the minimal kernel* of $\mathfrak{A}$ and will be denoted by $\mathfrak{A}^{\min}$. In other words, if $\mathfrak{A}$ is a Banach mapping ideal associated to the finitely-generated o.s. tensor norm α, then $\mathfrak{A}^{\min}$ will be the smallest Banach mapping ideal also associated to α.

In the Banach space framework, the representation theorem for minimal Banach ideals [23, 22.2] gives the connection with tensor products. In our setting, we will use this association to provide an o.s. structure to the mapping ideal $\mathfrak{A}^{\min}$.

Definition 9.1.1 Let $(\mathfrak{A}, \mathbf{A})$ be a mapping ideal. For $E, F \in \mathrm{ONORM}$ we define the class $\mathfrak{A}^{\min}(E, F)$ of all mappings $T : E \to F$ which admit a factorization of the form

$$
\begin{array}{ccc}
E & \xrightarrow{\ T\ } & F \\
{\scriptstyle R}\downarrow & & \uparrow{\scriptstyle S} \\
G & \xrightarrow[\ \widehat{T}\ ]{} & H,
\end{array}
\tag{9.1.1}
$$

J. A. Chávez-Domínguez et al., *Operator Space Tensor Norms*, Lecture Notes in Mathematics 2379, https://doi.org/10.1007/978-3-031-96209-7_9

where $G, H \in \text{OBAN}$, $\widehat{T} \in \mathfrak{A}(G, H)$ and R and S are approximable mappings. In this case we denote

$$\mathbf{A}^{\min}(T) := \inf\{\|R\|_{\mathrm{cb}}\mathbf{A}(\widehat{T})\|S\|_{\mathrm{cb}}\},$$

where the infimum runs over all the possible factorizations of T as above.

Example 9.1.2 It is clear that $\mathcal{F}^{\min} = \mathcal{F}$.

We now see that $\mathfrak{A}^{\min}(E, F)$ has certain structure.

Proposition 9.1.3 *Let* $(\mathfrak{A}, \mathbf{A})$ *be a Banach mapping ideal and* $E, F \in \text{ONORM}$. *Then the class* $\mathfrak{A}^{\min}(E, F)$ *is a vector space and* $\mathbf{A}^{\min}$ *is a norm. Moreover, if* $E, F \in$ *OBAN then* $(\mathfrak{A}^{\min}(E, F), \mathbf{A}^{\min})$ *is a Banach space.*

Proof It is clear that $\mathbf{A}^{\min}(T) \geq 0$, for all $T \in \mathfrak{A}^{\min}(E, F)$. Also, since for each factorization of T as in the definition $T = S \circ \widehat{T} \circ R$ we know that $\|\widehat{T}\|_{\mathrm{cb}} \leq \mathbf{A}(\widehat{T})$ and $\|T\|_{\mathrm{cb}} \leq \|R\|_{\mathrm{cb}}\|\widehat{T}\|_{\mathrm{cb}}\|S\|_{\mathrm{cb}}$ we derive that $\|T\|_{\mathrm{cb}} \leq \mathbf{A}^{\min}(T)$. Hence, $\mathbf{A}^{\min}(T) = 0$ implies $T = 0$.

Positive homogeneity is clear. Thus, to complete the first assertion we just need to check the triangle inequality. Let $T_1, T_2 \in \mathfrak{A}^{\min}(E, F)$. For a given $\varepsilon > 0$ there are factorizations

$$
\begin{array}{ccc}
E & \xrightarrow{T_1} & F \\
\scriptstyle R_1 \downarrow & & \uparrow \scriptstyle S_1 \\
G_1 & \xrightarrow[\widehat{T_1}]{} & H_1
\end{array}
\qquad\qquad
\begin{array}{ccc}
E & \xrightarrow{T_2} & F \\
\scriptstyle R_2 \downarrow & & \uparrow \scriptstyle S_2 \\
G_2 & \xrightarrow[\widehat{T_2}]{} & H_2
\end{array}
$$

with $R_j \in \mathcal{A}(E, G_j)$, $S_j \in \mathcal{A}(H_j, F)$, $\|R_j\|_{\mathrm{cb}} = \|S_j\|_{\mathrm{cb}} = 1$ and $\mathbf{A}(\widehat{T_j}) \leq (1 + \varepsilon)\mathbf{A}^{\min}(T_j)$, for $j = 1, 2$. Now, consider the factorization

$$
\begin{array}{ccc}
E & \xrightarrow{T_1 + T_2} & F \\
\scriptstyle R \downarrow & & \uparrow \scriptstyle S \\
G_1 \oplus_\infty G_2 & \xrightarrow[\widehat{T}]{} & H_1 \oplus_1 H_2,
\end{array}
$$

where $R(x) = (R_1(x), R_2(x))$, $\widehat{T}(g_1, g_2) = (\widehat{T_1}(g_1), \widehat{T_2}(g_2))$ and $S(h_1, h_2) = S_1(h_1) + S_2(h_2)$, for every $x \in E$, $g_j \in G_j$, $h_j \in H_j$, $j = 1, 2$.

Note that $\|R\|_{\mathrm{cb}} \leq \max\{\|R_1\|_{\mathrm{cb}}, \|R_2\|_{\mathrm{cb}}\} = 1$. By duality, the same argument yields $\|S\|_{\mathrm{cb}} = \|S'\|_{\mathrm{cb}} \leq \max\{\|S_1'\|_{\mathrm{cb}}, \|S_2'\|_{\mathrm{cb}}\} = 1$. Writing $\widehat{T} = i_1 \circ \widehat{T_1} \circ q_1 + i_2 \circ \widehat{T_2} \circ q_2$, where, for each $j = 1, 2$, $i_j : H_j \to H_1 \oplus_1 H_2$ is the canonical inclusion and $q_j : G_1 \oplus_\infty G_2 \to G_j$ is the canonical projection, we obtain $\mathbf{A}(\widehat{T}) \leq \mathbf{A}(\widehat{T_1}) + \mathbf{A}(\widehat{T_2})$.

In order to conclude $\mathbf{A}^{\min}(T_1 + T_2) \le (1 + \varepsilon)\left(\mathbf{A}^{\min}(T_1) + \mathbf{A}^{\min}(T_2)\right)$ we just need to check that $R \in \mathcal{A}(E, G_1 \oplus_\infty G_2)$ and $S \in \mathcal{A}(H_1 \oplus_1 H_2, F)$. The first one follows easily from the fact that R_1 and R_2 are approximable. The second one is derived by duality.

Now, it remains to prove that for $E, F \in$ OBAN, $\mathfrak{A}^{\min}(E, F)$ is complete. Let $\{T_k\}_k \subset \mathfrak{A}^{\min}(E, F)$ such that $\sum_{k=1}^\infty \mathbf{A}^{\min}(T_k) < \infty$. Take $(t_k)_k \subset \mathbb{R}$ with $1 \le t_k \uparrow \infty$ such that

$$\sum_{k=1}^\infty t_k \mathbf{A}^{\min}(T_k) \le (1 + \varepsilon) \sum_{k=1}^\infty \mathbf{A}^{\min}(T_k).$$

Note that the inequality $\|\cdot\|_{\mathrm{cb}} \le \mathbf{A}^{\min}$ tells us that a linear mapping $T = \sum_{k=1}^\infty T_k$ is well defined and that the series converges in $\mathrm{CB}(E, F)$.

For each k, take a factorization

$$
\begin{array}{ccc}
E & \xrightarrow{\ T_k\ } & F \\
R_k \downarrow & & \uparrow S_k \\
G_k & \xrightarrow[\ \widehat{T_k}\ ]{} & H_k
\end{array}
$$

with R_k, S_k approximable, $\|R_k\|_{\mathrm{cb}} = \|S_k\|_{\mathrm{cb}} = 1$ and $\mathbf{A}(\widehat{T_k}) \le (1 + \varepsilon)\mathbf{A}^{\min}(T_k)$. Then, consider

$$
\begin{array}{ccc}
E & \xrightarrow{\ T\ } & F \\
R \downarrow & & \uparrow S \\
c_0(G_k) & \xrightarrow[\ \widehat{T}\ ]{} & \ell_1(H_k),
\end{array}
$$

where $R(x) = \left(\frac{R_k(x)}{\sqrt{t_k}}\right)$, $\widehat{T}((g_k)_k) = \left((t_k \widehat{T_k}(g_k))_k\right)$ and $S((h_k)_k) = \sum_{k=1}^\infty \frac{S_k(h_k)}{\sqrt{t_k}}$, for every $x \in E$, $(g_k)_k \in c_0(G_k)$, $(h_k)_k \in \ell_1(H_k)$. It is easy to check that each mapping is well defined.

Let us see that $\widehat{T}$ belongs to $\mathfrak{A}(c_o(G_k), \ell_1(H_k))$. Note that, pointwise,

$$\widehat{T} = \sum_{j=1}^\infty t_j i_j \circ \widehat{T}_j \circ q_j, \tag{9.1.2}$$

where, for each j, $i_j : H_j \to \ell_1(H_k)$ is the canonical inclusion and $q_j : c_0(G_k) \to G_j$ is the canonical projection. Since

$$\sum_{j-1}^\infty \mathbf{A}(t_j i_j \circ \widehat{T}_j \circ q_j) \le \sum_{j=1}^\infty t_j \mathbf{A}(\widehat{T}_j) \le (1 + \varepsilon) \sum_{j=1}^\infty t_j \mathbf{A}^{\min}(T_j) < \infty$$

we obtain $\widehat{T} \in \mathfrak{A}(c_0(G_k), \ell_1(H_k))$. Moreover, the series (9.1.2) converges in $\mathfrak{A}$ and $\mathbf{A}(\widehat{T}) \leq \sum_{j=1}^{\infty} t_j \mathbf{A}(\widehat{T}_j)$.

Since $t_k \geq 1$ it is clear that $\|R : E \to c_0(G_k)\|_{\mathrm{cb}} \leq 1$. Now we prove that $R \in \mathcal{A}(E, c_0(G_k))$. For a given $\varepsilon > 0$ let $N \in \mathbb{N}$ such that $\frac{1}{\sqrt{t_N}} < \frac{\varepsilon}{2}$. For each $1 \leq j \leq N$, take $\widetilde{R}_j \in \mathcal{F}(E, G_j)$ satisfying $\|R_j - \widetilde{R}_j\|_{\mathrm{cb}} \leq \frac{\varepsilon}{2N}$. Define $\widetilde{R} : E \to c_0(G_k)$ by

$$x \mapsto \left(\frac{\widetilde{R}_1(x)}{\sqrt{t_1}}, \dots, \frac{\widetilde{R}_N(x)}{\sqrt{t_N}}, 0, 0, \dots \right).$$

Clearly, $\widetilde{R} \in \mathcal{F}(E, c_0(G_k))$ and $\|R - \widetilde{R}\|_{\mathrm{cb}} \leq \varepsilon$. Therefore, $R \in \mathcal{A}(E, c_0(G_k))$ with $\|R\|_{\mathrm{cb}} \leq 1$.

Note that $S' : F' \to \ell_\infty(H_k')$ is given by $S'(y') = (\frac{S_k'(y')}{\sqrt{t_k}})_k$, so the same argument as above yields $\|S\|_{\mathrm{cb}} = \|S'\|_{\mathrm{cb}} \leq 1$. We finally check that $S \in \mathcal{A}(\ell_1(H_k), F)$. Also, analogously as before we can prove that for each $\varepsilon > 0$ there is $\widetilde{S} \in \mathcal{F}(\ell_1(H_k), F)$ such that $\|S - \widetilde{S}\|_{\mathrm{cb}} \leq \varepsilon$.

From the arguments above we certainly obtained that $T \in \mathfrak{A}^{\min}(E, F)$ with $\mathbf{A}^{\min}(T) \leq \sum_{k=1}^{\infty} \mathbf{A}^{\min}(T_k)$. This implies that $T = \sum_{k=1}^{\infty} T_k$ converges in $\mathfrak{A}^{\min}(E, F)$, which finishes the proof. $\qquad\square$

9.2 The Representation Theorem

In order to relate the tensor product $E' \otimes_\alpha F$ with $\mathfrak{A}^{\min}(E, F)$ as in the Banach space setting, we need the following lemma.

Lemma 9.2.1 *Let $(\mathfrak{A}, \mathbf{A})$ be a Banach mapping ideal associated to the finitely-generated o.s. tensor norm α. Then for any $E, F \in \mathrm{ONORM}$ the natural map*

$$J : E' \otimes_\alpha F \to \mathfrak{A}^{\min}(E, F) \tag{9.2.1}$$

is a contraction of normed spaces, i.e., $\mathbf{A}^{\min}(J(z)) \leq \alpha(z; E', F)$, $\forall z \in E' \otimes F$.

Proof Let $z \in E' \otimes F$, $M \in \mathrm{OFIN}(E')$, $N \in \mathrm{OFIN}(F)$, $u \in M \otimes N$ with $z = (i_M^{E'} \otimes i_N^F)(u)$.

From the diagram

$$
\begin{array}{ccccc}
E' \otimes_\alpha F & \xrightarrow{\ J\ } & \mathfrak{A}^{\min}(E, F) & \qquad & i_N^F \circ T \circ q_{\circ M}^E \\[2mm]
\big\uparrow{\scriptstyle i_M^{E'} \otimes i_N^F} & & \big\uparrow & & \wedge \\[2mm]
M \otimes_\alpha N & =\!=\!= & \mathfrak{A}(E/{}^\circ M, N) & & T
\end{array}
$$

we get

$$\mathbf{A}^{\min}(J(z)) = \mathbf{A}^{\min}(i_N^F \circ T_u \circ q_{\circ M}^E) \leq \mathbf{A}(T_u) = \alpha(u; M, N).$$

Since α is finitely-generated we obtain $\mathbf{A}^{\min}(J(z)) \leq \alpha(z; E', F)$. $\square$

A canonical argument shows that each mapping in $\mathfrak{A}^{\min}$ can be approximated (in the norm of the ideal) by an element from $\mathfrak{A}$ composed at both sides with finite-rank mappings.

Lemma 9.2.2 *Let* $T \in \mathfrak{A}^{\min}(E, F)$ *with factorization* $T = S \circ \widehat{T} \circ R$ *as in* (9.1.1). *Then, there are sequences* $(R_n)_n \subset \mathcal{F}(E, G)$ *and* $(S_n)_n \subset \mathcal{F}(H, F)$ *such that* $S_n \circ \widehat{T} \circ R_n \to T$ *in* $\mathfrak{A}^{\min}(E, F)$.

Proof We can take $(R_n)_n \subset \mathcal{F}(E, G)$ and $(S_n)_n \subset \mathcal{F}(H, F)$ converging to R and S, respectively, in the cb-norm. Then,

$$T - S_n \circ \widehat{T} \circ R_n = (S_n - S) \circ \widehat{T} \circ R_n + S \circ \widehat{T} \circ (R - R_n).$$

Now, a standard computation using the definition of the $\mathbf{A}^{\min}$ norm, yields the result.
$\square$

We are now ready to prove one of the main results of this chapter, which will allow us to provide an o.s. structure for $\mathfrak{A}^{\min}$.

Theorem 9.2.3 *Let* $(\mathfrak{A}, \mathbf{A})$ *be a Banach mapping ideal associated to the finitely-generated o.s. tensor norm* α. *Then for any* $E, F \in \mathrm{OBAN}$ *the natural map*

$$J : E' \widehat{\otimes}_\alpha F \to \mathfrak{A}^{\min}(E, F)$$

is a metric surjection (at the Banach space level).

Proof Using Lemma 9.2.1, we know that J is a contraction. Take $R \in \mathcal{F}(E, G)$, $\widehat{T} \in \mathfrak{A}(G, H)$ and $S \in \mathcal{F}(H, F)$. If $z \in E' \otimes F$ satisfies $J(z) = S \circ \widehat{T} \circ R$ consider the factorization

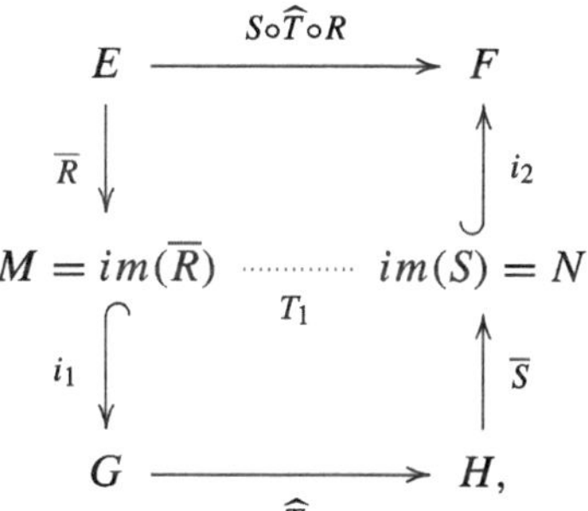

where $\overline{R}$ and $\overline{S}$ are the canonical restrictions to their images. Thus, since M, $N \in$ OFIN,

$$\alpha(z; E', F) = \alpha((\overline{R}' \otimes i_2)z_{T_1}; E', F) \le \|\overline{R}'\|_{\mathrm{cb}}\|i_2\|_{\mathrm{cb}}\alpha(z_{T_1}; M', N)$$
$$= \|\overline{R}\|_{\mathrm{cb}}\mathbf{A}(T_1) \le \|\overline{R}\|_{\mathrm{cb}}\|i_1\|_{\mathrm{cb}}\mathbf{A}(\widehat{T})\|\overline{S}\|_{\mathrm{cb}} \qquad (9.2.2)$$
$$= \|R\|_{\mathrm{cb}}\mathbf{A}(\widehat{T})\|S\|_{\mathrm{cb}}.$$

Now, let $T \in \mathfrak{A}^{\min}(E, F)$. Given $\varepsilon > 0$, find a factorization $T = S \circ \widehat{T} \circ R$ with $R \in \mathcal{A}(E, G)$, $S \in \mathcal{A}(H, F)$ and $\|R\|_{\mathrm{cb}}\mathbf{A}(\widehat{T})\|S\|_{\mathrm{cb}} \le (1 + \varepsilon)\mathbf{A}^{\min}(T)$. By Lemma 9.2.2, we know that there are sequences $(R_n)_{n\in\mathbb{N}} \subset \mathcal{F}(E, G)$ and $(S_n)_{n\in\mathbb{N}} \subset \mathcal{F}(H, F)$ such that $\|R_n - R\|_{\mathrm{cb}} \to 0$, $\|S_n - S\|_{\mathrm{cb}} \to 0$ and thus $S_n \circ \widehat{T} \circ R_n$ converges to T in $\mathfrak{A}^{\min}(E, F)$. Now, pick $z_n \in E' \otimes F$ with $J(z_n) = S_n \circ \widehat{T} \circ R_n$. Then,

$$J(z_n - z_m) = J(z_n) - J(z_m)$$
$$= S_n \circ \widehat{T} \circ R_n - S_m \circ \widehat{T} \circ R_m$$
$$= (S_n - S_m) \circ \widehat{T} \circ R_n + S_m \circ \widehat{T} \circ (R_n - R_m)$$

By reasoning as in (9.2.2) and using the triangle inequality we have

$$\alpha(z_n - z_m; E', F) \le \|S_n - S_m\|_{\mathrm{cb}}\mathbf{A}(\widehat{T})\|R_n\|_{\mathrm{cb}} + \|S_m\|_{\mathrm{cb}}\mathbf{A}(\widehat{T})\|R_n - R_m\|_{\mathrm{cb}}.$$

Therefore, $(z_n)_{n\in\mathbb{N}}$ is a Cauchy sequence in $E'\widehat{\otimes}_\alpha F$ with limit, say, z. since J is continuous,

$$J(z) = \lim_{n\to\infty} J(z_n) = \lim_{n\to\infty} S_n \circ \widehat{T} \circ R_n,$$

where the limits are in $\mathfrak{A}^{\min}(E, F)$. We have checked above that the latter limit is in fact T. In other words, given $T \in \mathfrak{A}^{\min}(E, F)$ we have found $z \in E'\widehat{\otimes}_\alpha F$ such that $J(z) = T$ and, moreover, by (9.2.2) again

$$\alpha(z; E', F) = \lim_{n\to\infty} \alpha(z_n; E', F)$$
$$\le \lim_{n\to\infty} \|S_n\|_{\mathrm{cb}}\mathbf{A}(\widehat{T})\|R_n\|_{\mathrm{cb}}$$
$$\le \mathbf{A}^{\min}(T)(1 + \varepsilon).$$

This shows that the mapping J is a metric surjection (since we already knew that $\|J\| \le 1$). $\square$

Using the previous theorem we can now provide an operator space structure for $\mathfrak{A}^{\min}(E, F)$.

Definition 9.2.4 Given a Banach mapping ideal $(\mathfrak{A}, \mathbf{A})$, we endow $\mathfrak{A}^{\min}(E, F)$ with the unique o.s. structure such that the natural map J becomes a complete quotient mapping. Namely, given $T \in M_n(\mathfrak{A}^{\min}(E, F))$, we define

$$\mathbf{A}_n^{\min}(T) := \inf \alpha_n(z; E', F),$$

where the infimum runs all over $z \in M_n(E' \widehat{\otimes} F)$ such that $J_n(z) = T$.

With this structure we have obviously the following:

Theorem 9.2.5 (Representation Theorem for Minimal Mapping Ideals) *Let*
$(\mathfrak{A}, \mathbf{A})$ be a Banach mapping ideal associated to the finitely-generated o.s. tensor
norm α. Then for any $E, F \in$ OBAN the natural operator

$$J : E' \widehat{\otimes}_\alpha F \longrightarrow \mathfrak{A}^{\min}(E, F) \tag{9.2.3}$$

is a complete quotient mapping. In particular, if $M, N \in$ OFIN then we have a
complete isometry $\mathfrak{A}^{\min}(M, N) = M' \widehat{\otimes}_\alpha N$.

Theorem 9.2.6 *Let $(\mathfrak{A}, \mathbf{A})$ be a Banach mapping ideal associated to the finitely-*
generated o.s. tensor norm α. Then $(\mathfrak{A}^{\min}, \mathbf{A}^{\min})$ is also a Banach mapping ideal
associated to α. Moreover, $\mathfrak{A}^{\min}$ is the smallest Banach mapping ideal associated
to α.

Proof To see that $(\mathfrak{A}^{\min}, \mathbf{A}^{\min})$ is a mapping ideal, we first see that the identity map $\mathfrak{A}^{\min}(E, F) \to \mathrm{CB}(E, F)$ is a complete contraction. Consider the following commutative diagram:

$$
\begin{array}{ccc}
E' \widehat{\otimes}_\alpha F & \xrightarrow{\ \ J\ \ } & \mathfrak{A}^{\min}(E, F) \\
\Big\downarrow & & \Big\downarrow \\
E' \widehat{\otimes}_{\min} F & \xhookrightarrow{\ \ i\ \ } & \mathrm{CB}(E, F),
\end{array}
\tag{9.2.4}
$$

Now, by the Representation theorem for minimal mapping ideals we know that J is complete quotient. On the other hand, the inclusion i is a complete isometry (see Remark 5.2.2) and the canonical mapping $E' \widehat{\otimes}_\alpha F \to E' \widehat{\otimes}_{\min} F$ is a complete contraction. By simply following the diagram, this implies that the descending arrow $\mathfrak{A}^{\min}(E, F) \dashrightarrow \mathrm{CB}(E, F)$ is a complete contraction.

Note that condition (b') immediately follows from the last assertion of the previous theorem: $\mathfrak{A}^{\min}(M, N) = M' \widehat{\otimes}_\alpha N$.

Let us now focus on the ideal property. Consider $T \in M_n(\mathfrak{A}^{\min}(E, F))$, $r \in \mathrm{CB}(E_0, E)$ and $s \in \mathrm{CB}(F, F_0)$. By the ideal property of $\mathcal{A}$ it is easy to see that $s_n \circ T \circ r$ lies in $M_n(\mathfrak{A}^{\min}(E_0, F_0))$. To bound the norm consider the following commutative diagram.

$$
\begin{array}{ccc}
E'\widehat{\otimes}_\alpha F \longrightarrow \mathfrak{A}^{\min}(E,F) & \qquad & T \\
{\scriptstyle r'\otimes s}\Big\downarrow \qquad\qquad \Big\downarrow & & \Big\downarrow \\
E_0'\widehat{\otimes}_\alpha F_0 \longrightarrow \mathfrak{A}^{\min}(E_0,F_0) & \qquad & s\circ T\circ r.
\end{array}
\qquad (9.2.5)
$$

Since the horizontal arrows are complete quotients and the mapping

$$
r'\otimes s : E'\widehat{\otimes}_\alpha F \to E_0'\widehat{\otimes}_\alpha F_0
$$

has cb-norm bounded by $\|s\|_{\mathrm{cb}}\|r\|_{\mathrm{cb}}$, it follows that

$$
\mathbf{A}_n^{\min}(s_n \circ T \circ r) \le \|s\|_{\mathrm{cb}}\,\mathbf{A}_n^{\min}(T)\,\|r\|_{\mathrm{cb}}\,.
$$

Appealing again to the complete isometry $\mathfrak{A}^{\min}(M,N) = M'\widehat{\otimes}_\alpha N$, valid for $M, N \in \mathrm{OFIN}$, we derive that the mapping ideal $\mathfrak{A}^{\min}$ is associated to α.

Finally, we show that $\mathfrak{A}^{\min}$ is the smallest Banach mapping ideal associated to α. Since for any Banach mapping ideal $\mathfrak{B}$ we have $\mathfrak{B}^{\min} \subset \mathfrak{B}$ it is enough to see that if $\mathfrak{B}$ is also associated to α then $\mathfrak{A}^{\min} = \mathfrak{B}^{\min}$. Indeed, this is straightforward since we have the complete quotients $J : E'\widehat{\otimes}_\alpha F \twoheadrightarrow \mathfrak{A}^{\min}(E,F)$ and $J : E'\widehat{\otimes}_\alpha F \twoheadrightarrow \mathfrak{B}^{\min}(E,F)$. $\qquad\square$

Now we present an operator space version of [23, Prop. 22.1].

Proposition 9.2.7 *Let $(\mathfrak{A}, \mathbf{A})$ be a Banach mapping ideal associated to the finitely-generated o.s. tensor norm α. Then,*

(1) $\mathfrak{A}^{\min}(M,N) = \mathfrak{A}^{\max}(M,N) = \mathfrak{A}(M,N) = M'\otimes_\alpha N$ holds completely isometrically for every $M, N \in \mathrm{OFIN}$.

(2) $\mathfrak{A}^{\min\,\max} = \mathfrak{A}^{\max}$.

(3) $\mathfrak{A}^{\max\,\min} = \mathfrak{A}^{\min}$.

(4) Given $E, F \in \mathrm{OBAN}$ and $T \in M_n(\mathfrak{A}^{\min}(E,F))$ there are $T_k \in M_n(\mathcal{F}(E,F))$ such that $\mathbf{A}_n^{\min}(T_k - T) \to 0$.

(5) Let $\mathfrak{B}$ be a Banach mapping ideal. Suppose that for every $M, N \in \mathrm{OFIN}$ we have that the inclusion $\|\mathfrak{A}(M,N) \hookrightarrow \mathfrak{B}(M,N)\|_{\mathrm{cb}} \le c$ then $\mathfrak{A}^{\min} \subset \mathfrak{B}^{\min}$ and $\|\mathfrak{A}^{\min}(E,F) \hookrightarrow \mathfrak{B}^{\min}(E,F)\|_{\mathrm{cb}} \le c$ for every $E, F \in \mathrm{OBAN}$.

Proof (1), (2) and (3) follow directly from the Representation Theorems for maximal and minimal ideals.

(4) This can easily be derived from Lemma 9.2.2.

(5) Let β be the finitely-generated o.s. tensor norm associated to $\mathfrak{B}$. The hypothesis translates into $\|M'\widehat{\otimes}_\alpha N \hookrightarrow M'\widehat{\otimes}_\beta N\|_{\mathrm{cb}} \le c$ for every $M, N \in \mathrm{OFIN}$. Since both norms are finitely-generated, this completely bounded inclusion can be extended to OBAN. Also, by Proposition 8.2.6, we know that $\mathfrak{A}^{\max} \subset \mathfrak{B}^{\max}$. Now, using (3), we have $\mathfrak{A}^{\min} \subset \mathfrak{B}^{\min}$. To conclude, consider the following diagram

$$E' \widehat{\otimes}_\alpha \Gamma \xrightarrow{\ J_\alpha\ } \mathfrak{A}^{\min}(E, F)$$

$$\iota \downarrow \qquad\qquad\qquad \downarrow$$

$$E' \widehat{\otimes}_\beta F \xrightarrow{\ J_\beta\ } \mathfrak{B}^{\min}(E, F). \tag{9.2.6}$$

The result follows from the fact that J_α and J_β are complete quotients and $\|\iota\|_{\mathrm{cb}} \leq c$.
$\square$

Definition 9.2.8 We say that a Banach mapping ideal $(\mathfrak{A}, \mathbf{A})$ is minimal whenever $(\mathfrak{A}, \mathbf{A}) = (\mathfrak{A}^{\min}, \mathbf{A}^{\min})$.

We now provide some examples

(i) $\mathcal{A} = \mathrm{CB}^{\min}$. Moreover, given $E, F \in \mathrm{OBAN}$ we have the completely isometric isomorphism $\mathcal{A}(E, F) = E' \widehat{\otimes}_{\min} F$.

   Indeed, it is known that the canonical mapping $i : E' \widehat{\otimes}_{\min} F \hookrightarrow \mathrm{CB}(E, F)$ is completely isometric and by the Representation Theorem, $J : E' \widehat{\otimes}_{\min} F \twoheadrightarrow \mathrm{CB}^{\min}(E, F)$ is a complete quotient. Then, the inclusion $\mathrm{CB}^{\min}(E, F) \hookrightarrow \mathrm{CB}(E, F)$ has to be completely isometric. That means that the norm in $\mathrm{CB}^{\min}(E, F)$ is $\| \cdot \|_{\mathrm{cb}}$. Since $\mathcal{A}$ is contained in any Banach mapping ideal whose norm is the cb-norm, it should be $\mathcal{A} = \mathrm{CB}^{\min}$. Also, J is a completely isometric isomorphism and so $\mathcal{A}(E, F) = E' \widehat{\otimes}_{\min} F$.

(ii) $\mathcal{N} = \mathcal{I}^{\min}$.

   Indeed, since $\mathcal{I} \sim \mathrm{proj}$ we have a complete quotient $J : E' \widehat{\otimes}_{\mathrm{proj}} F \twoheadrightarrow \mathcal{I}^{\min}(E, F) \subset \mathrm{CB}(E, F)$ and this coincides with the definition of $\mathcal{N}(E, F)$.

(iii) The ideal of completely right-p nuclear mappings $\mathcal{N}^p$ is minimal.

   Indeed, it is clear from its definition that $\mathcal{N}^p = (\mathcal{N}^p)^{\min}$ because $\mathcal{N}^p \sim d_p$ and $J^p : E' \widehat{\otimes}_{d_p} F \twoheadrightarrow \mathcal{N}^p(E, F) \subset \mathrm{CB}(E, F)$ is a complete quotient.

Chapter 10
Completely Projective/Injective Operator Space Tensor Norms

In this chapter we return to the abstract theory of o.s. tensor norms. In particular, we study o.s. tensor norms which behave well with respect to complete injections or projections. We know that the minimal o.s. tensor norm respects complete injections [34, Prop. 8.1.5] whereas the projective o.s. tensor norm respects complete projections [34, Prop. 7.1.7]. A well-known property of the Haagerup o.s. tensor norm is that it respects both complete projections and injections [34, Prop. 9.2.5]. It should be noted that this cannot happen in the Banach space/normed space framework as a consequence of Grothendieck's inequality [23, Prop. 20.20]. Based on the classical definitions, we consider in this chapter two natural procedures on o.s. tensor norms: the completely injective hull and the completely projective hull. Similar constructions were considered in [10].

10.1 Injectivity and Projectivity

We begin by recalling from Sect. 2.1.1 the definitions of completely injective/projective o.s. tensor norms.

Definition 10.1.1 An o.s. tensor norm α on ONORM is called *completely right-injective* if for all $E, F, G \in$ ONORM and for all complete injections $i : F \hookrightarrow G$ the mapping

$$id_E \otimes i : E \otimes_\alpha F \to E \otimes_\alpha G$$

is a complete injection and it is called *completely right-projective* if for all $E, F, G \in$ ONORM and for all complete metric surjections $q : G \twoheadrightarrow F$, and all spaces E the mapping

$$id_E \otimes q : E \otimes_\alpha G \to E \otimes_\alpha F$$

J. A. Chávez-Domínguez et al., *Operator Space Tensor Norms*, Lecture Notes in Mathematics 2379, https://doi.org/10.1007/978-3-031-96209-7_10

is a complete metric surjection. Left versions are defined analogously, and we say that α is *completely injective* (resp. *completely projective*) when it is both completely left- and completely right-injective (resp. projective). Analogous definitions will be used for other classes of spaces besides ONORM (OFIN, for example).

To provide some examples of these definitions, in [18, Prop. 3.4] it is shown that the Chevet-Saphar tensor norms d_p are completely right-projective and the Chevet-Saphar tensor norms g_p are completely left-projective. Also, by [64, Prop. 6.1], the norm d_2 is completely left-injective and g_2 is completely right-injective. It should also be mentioned that, by [75, Prop. 6.2], any λ-o.s. tensor norm is completely projective. For some conditions implying that a λ-o.s. tensor norm is completely injective, see [2, Prop. 1.2].

For $M, N \in$ OFIN, the complete isometry

$$M \otimes_{\alpha'} N = (M' \otimes_\alpha N')'$$

implies the following:

Remark 10.1.2 α is completely right-projective on OFIN if and only if α' is completely right-injective on OFIN.

In the Banach space world, the version of this relationship is valid not only in the finite-dimensional case. Surprisingly, the situation is not the same in our framework: this property cannot be extended into OBAN for general o.s. tensor norms as it will be shown in Remark 12.1.5 (c). This is an important difference between the classical and the non-commutative theory of tensor norms. Namely, in OBAN, the fact that an o.s. tensor norm is completely right-injective does not imply that its dual norm is completely right-projective. Nevertheless, the converse is valid (see Corollary 10.1.4 below).

Proposition 10.1.3 *If α is finitely-generated and completely right-injective on* OFIN *then α is completely right-injective on* ONORM.

Proof Let $i : F \hookrightarrow G$ be a complete injection and $z \in M_n(E \otimes F)$. Since α is finitely-generated, given $\varepsilon > 0$, consider $M \in$ OFIN(E) and $N \in$ OFIN(G) such that

$$\alpha_n((id_E \otimes i)_n(z); M, N) \leq (1 + \varepsilon)\alpha_n((id_E \otimes i)_n(z); E, G).$$

Given $\widetilde{M} \in$ OFIN(E) and $\widetilde{N} \in$ OFIN(F) such that $z \in M_n(\widetilde{M} \otimes \widetilde{N})$ we have:

$$\alpha_n(z; E, F) \leq \alpha_n(z; M + \widetilde{M}, \widetilde{N}) = \alpha_n((id_E \otimes i)_n(z); M + \widetilde{M}, N + i(\widetilde{N}))$$

$$= \alpha_n((id_E \otimes i)_n(z); M, N) \leq (1 + \varepsilon)\alpha_n((id_E \otimes i)_n(z); E, G),$$

where equalities are due to the fact that α is completely right-injective on OFIN. This concludes the proof since we always have $\alpha_n((id_E \otimes i)_n(z); E, G) \leq \alpha_n(z; E, F)$. $\qquad\square$

As a consequence of Remark 10.1.2, Proposition 10.1.3 and the fact that α' is finitely-generated we have:

Corollary 10.1.4 *If α is completely right-projective on* OFIN *then α' is completely right-injective on* ONORM.

As mentioned above, the reciprocal of the previous statement does not hold in general. Nevertheless, under an additional hypothesis on the tensor norm we can actually get the converse (see Proposition 11.1.1).

Remark 10.1.5 It is clear from the definitions that if α and β are completely right-injective o.s. tensor norms then it is so $\alpha \cap \beta$. Also, if α and β are completely right-projective o.s. tensor norms then $\alpha + \beta$ is completely right-projective too.

On the other hand, the intersection procedure does not preserve projectivity and the sum does not preserve injectivity. We have already commented in Example 2.2.2 that [53, p. 279] proves that $h+h^t$ is not completely injective, but in fact it shows that it is not right-injective. Specifically, what is shown is that there exists a completely isometric embedding $i : E \to F$ between finite-dimensional operator spaces, and a tensor $u \in E' \otimes E$ such that

$$(h + h^t)\big((Id_{E'} \otimes i)u; E', F\big) < (h + h^t)(u; E', E).$$

As a consequence, due to Corollary 10.1.4 and Proposition 5.1.3, $h \cap h^t$ could not be completely right-projective.

10.2 Completely Injective Hulls

As in the Banach space setting we will now describe an analogous theory for the injective hulls of a given o.s. tensor norm.

Theorem 10.2.1 *Let α be an o.s. tensor norm on* ONORM. *Then there is a unique completely right-injective o.s. tensor norm $\alpha\backslash$ on* ONORM *such that $\beta \leq \alpha\backslash$ for all completely right-injective o.s. tensor norms $\beta \leq \alpha$. For all normed operator spaces E and $F \subseteq \mathcal{B}(H)$,*

$$E \otimes_{\alpha\backslash} F \to E \otimes_\alpha \mathcal{B}(H)$$

is a complete isometry.

Proof Suppose that we have two completely isometric embeddings $i_j : F \to \mathcal{B}(H_j)$, $j = 1, 2$ and define operator space structures α_j on $E \otimes F$ induced by

$$id_E \otimes i_j : E \otimes F \to E \otimes_\alpha \mathcal{B}(H_j).$$

By the injectivity of $\mathcal{B}(H_j)$, there exist completely bounded contractions $R_1 : \mathcal{B}(H_1) \to \mathcal{B}(H_2)$ and $R_2 : \mathcal{B}(H_2) \to \mathcal{B}(H_1)$ such that $i_1 = R_2 \circ i_2$ and $i_2 = R_1 \circ i_1$.

The map $(id_E \otimes R_1) \circ (id_E \otimes i_1)$ shows that the identity $E \otimes_{\alpha_1} F \to E \otimes_{\alpha_2} F$ is completely contractive, and analogously so is $E \otimes_{\alpha_2} F \to E \otimes_{\alpha_1} F$. Therefore, there is no ambiguity if we define $\alpha\backslash$ on $E \otimes F$ to be the operator space structure on $E \otimes F$ induced by the particular embedding

$$E \otimes F \to E \otimes_\alpha \mathcal{B}(H).$$

From the injectivity of min, the fact that $\min \leq \alpha$ and the diagram

$$
\begin{array}{ccc}
E \otimes_{\alpha\backslash} F & \longrightarrow & E \otimes_\alpha \mathcal{B}(H) \\
\downarrow & & \downarrow \\
E \otimes_{\min} F & \longrightarrow & E \otimes_{\min} \mathcal{B}(H)
\end{array}
$$

it follows that $\min \leq \alpha\backslash$. Similarly, from $\alpha \leq \mathrm{proj}$ and

$$
\begin{array}{ccc}
E \otimes_{\mathrm{proj}} F & \longrightarrow & E \otimes_{\mathrm{proj}} \mathcal{B}(H) \\
\downarrow & & \downarrow \\
E \otimes_{\alpha\backslash} F & \longrightarrow & E \otimes_\alpha \mathcal{B}(H)
\end{array}
$$

it follows that $\alpha\backslash \leq \mathrm{proj}$. Thus, it is clear that $\alpha\backslash$ is a reasonable operator space cross-norm.

It is not hard to see that $\alpha\backslash$ has the complete metric mapping property. Let $S \in \mathrm{CB}(E_1, E_2)$ and $T \in \mathrm{CB}(F_1, F_2)$ and consider completely isometric embeddings $F_j \subseteq \mathcal{B}(K_j)$. By the injectivity of $\mathcal{B}(H)$, there exists a completely bounded map $\widehat{T} : \mathcal{B}(K_1) \to \mathcal{B}(K_2)$ such that $\|\widehat{T}\|_{\mathrm{cb}} = \|T\|_{\mathrm{cb}}$ and

$$
\begin{array}{ccc}
\mathcal{B}(K_1) & \xrightarrow{\;\widehat{T}\;} & \mathcal{B}(K_2) \\
\uparrow & & \uparrow \\
F_1 & \xrightarrow{\;T\;} & F_2
\end{array}
$$

commutes. After tensorizing,

$$
\begin{array}{ccc}
E_1 \otimes_\alpha \mathcal{B}(K_1) & \xrightarrow{\;S \otimes \widehat{T}\;} & E_2 \otimes_\alpha \mathcal{B}(K_2) \\
\uparrow & & \uparrow \\
E_1 \otimes_{\alpha\backslash} F_1 & \xrightarrow{\;S \otimes T\;} & E_2 \otimes_{\alpha\backslash} F_2
\end{array}
$$

from where we clearly see that, by the complete metric mapping property of α,

$$\left\| S \otimes T : E_1 \otimes_{\alpha\backslash} F_1 \to E_2 \otimes_{\alpha\backslash} F_2 \right\|_{\mathrm{cb}} \leq \|S\|_{\mathrm{cb}} \|T\|_{\mathrm{cb}} .$$

Suppose that $i : F \to G$ and $i_G : G \to \mathcal{B}(H)$ are complete injections. Since we had already shown that $\alpha\backslash$ does not depend on the particular embedding into a $\mathcal{B}(H)$ and $i_G \circ i : F \to \mathcal{B}(H)$ is a complete injection we have, for any $n \in \mathbb{N}$ and $z \in M_n(E \otimes F)$,

$$\alpha\backslash_n((id_E \otimes i)_n(z); E, G) \leq \alpha\backslash_n(z; E, F) = \alpha_n((id_E \otimes i_G \circ i)_n(z); E, \mathcal{B}(H))$$

$$= \alpha\backslash_n((id_E \otimes i)_n(z); E, G),$$

showing that $\alpha\backslash$ is completely-right injective.

Finally, if β is a completely right-injective o.s. tensor norm dominated by α then the diagram

$$
\begin{array}{ccc}
E \otimes_{\alpha\backslash} F & \longrightarrow & E \otimes_{\alpha} \mathcal{B}(H) \\
\Big\downarrow & & \Big\downarrow \\
E \otimes_{\beta} F & \longrightarrow & E \otimes_{\beta} \mathcal{B}(H)
\end{array}
$$

shows $\beta \leq \alpha\backslash$. $\qquad\qquad\square$

Lemma 10.2.2 *Let α be a finitely-generated o.s. tensor norm then $\alpha\backslash$ is also finitely-generated.*

Proof Let E and F in ONORM and $z \in M_n(E \otimes F)$. By the definition of the o.s. tensor norm $\alpha\backslash$ if $F \subseteq \mathcal{B}(H)$ we have

$$(\alpha\backslash)_n(z; E, F) = \alpha_n(z; E, \mathcal{B}(H)).$$

Since α is finitely-generated, given $\varepsilon > 0$ there are $\widetilde{M} \in \mathrm{OFIN}(E)$ and $\widetilde{N} \in \mathrm{OFIN}(\mathcal{B}(H))$ such that

$$\alpha_n(z; \widetilde{M}, \widetilde{N}) \leq (1 + \varepsilon)\alpha_n(z; E, \mathcal{B}(H)).$$

Now, given $M \in \mathrm{OFIN}(E)$ and $N \in \mathrm{OFIN}(F)$ such that $z \in M \otimes N$ we have that

$$(\alpha\backslash)_n(z; (M + \widetilde{M}), N) = (\alpha\backslash)_n(z; (M + \widetilde{M}), (N + \widetilde{N}))$$

$$\leq \alpha_n(z; (M + \widetilde{M}), (N + \widetilde{N})) \leq \alpha_n(z; \widetilde{M}, \widetilde{N})$$

$$\leq (1 + \varepsilon)\alpha_n(z; E, \mathcal{B}(H)) = (1 + \varepsilon)(\alpha\backslash)_n(z; E, F).$$

This concludes the proof. $\qquad\qquad\square$

Definition 10.2.3 The o.s. tensor norm $\alpha\backslash$ given in Theorem 10.2.1 is called the *completely right-injective hull* of α. One can define analogously the *completely left-injective hull* of α, denoted $/\alpha$, and the *completely injective hull*

$$/\alpha\backslash := (/\alpha)\backslash = /(\alpha\backslash).$$

Note that there is no ambiguity in the previous equality since $(/\alpha)\backslash = /(\alpha\backslash)$. Indeed, let $E, F \in$ ONORM and suppose that $E \subset \mathcal{B}(H_E)$ and $F \subset \mathcal{B}(H_F)$; by Theorem 10.2.1 and its left version, the o.s. tensor norms $(/\alpha)\backslash$ and $/(\alpha\backslash)$ can be computed on $E \otimes F$ through the complete isometry

$$E \otimes F \hookrightarrow \mathcal{B}(H_E) \otimes_\alpha \mathcal{B}(H_F).$$

10.3   Completely Projective Hulls

Now it is time to describe the completely projective hulls of a given o.s. tensor norm. To do this we need some preliminary results that will be useful later. Let us start by proving an operator space version of the perturbation step in the proof of [23, Lem. 20.2.(2)] without appealing to local reflexivity. The proof is inspired by [65, Lem. 2.13.2].

Lemma 10.3.1 *Let $F \in$ ONORM, and let $\overline{F}$ be its completion. Given $N \in$ OFIN$(\overline{F})$ and $\varepsilon > 0$, there exists a linear map $R : N \to F$ with $\|R\|_{\mathrm{cb}} \le 1 + \varepsilon$ and $Ry = y$ for all $y \in F \cap N$.*

Proof Let $(x_j, x'_j)_{j=1}^m$ be a biorthogonal system for the space N, that is, a basis $x_1, \ldots, x_m$ of N and functionals $x'_1, \ldots, x'_m \in N'$ satisfying $x'_k(x_j) = \delta_{kj}$ (we don't assume anything about their norms). Without loss of generality, we may suppose that $\{x_j\}_{j=1}^{m_0}$ is a basis for $F \cap N$ (with $m_0 = 0$ if $F \cap N = \{0\}$). For $1 \le j \le m_0$, let $y_j = x_j$, and for $m_0 + 1 \le j \le m$, choose $y_j \in F$ close enough to x_j so that

$$\sum_{j=m_0+1}^m \left\| x'_j \right\| \left\| x_j - y_j \right\| < \varepsilon.$$

Define $R : N \to F$ as $y \mapsto \sum_{j=1}^m x'_j(y)y_j$. Since the identity map $I : N \to \overline{F}$ can be written as $y \mapsto \sum_{j=1}^m x'_j(y)x_j$, it is clear that $Ry = y$ for all $y \in F \cap N$, and

$$\|R\|_{\mathrm{cb}} \le \|I\|_{\mathrm{cb}} + \sum_{j=m_0+1}^m \left\| x'_j \right\| \left\| x_j - y_j \right\| < 1 + \varepsilon.$$

This concludes the proof.   □

Let us state an important Quotient lemma regarding metric surjections from the classical theory [23, Lem. 7.4]. This can be translated to the operator space setting canonically adapting the proof (note that part of the argument is explained in the proof of Proposition 2.2.4 (iii)). Its statement is the following:

Lemma 10.3.2 *Let $E, F \in$ ONORM, $E_0 \subset E$ a dense subspace, $q \in CB(E, F)$ a surjective mapping and consider $q_0 = q|_{E_0} : E_0 \to q(E_0)$ its restriction. Then, q_0 is a complete metric surjection if and only if $\overline{\ker q_0} = \ker q$ and q is a complete metric surjection.*

Another ingredient needed for the construction of the completely projective hull is the following concept of right-finite hull of an o.s. tensor norm:

Definition 10.3.3 Given normed operator spaces E and F, an o.s. tensor norm α on OFIN and $u \in M_n(E \otimes F)$, the *right-finite hull* of α is given by

$$\alpha_n^{\to}(u; E, F) = \inf \{\alpha_n(u; E, F_0) : F_0 \in \text{OFIN}(F), u \in M_n(E \otimes F_0)\}.$$

It is clear that $\alpha^{\to}$ is an o.s. tensor norm and $\alpha \leq \alpha^{\to} \leq \overrightarrow{\alpha}$.

We are now ready to present the operator space version of [23, Lem. 20.2]:

Lemma 10.3.4 *Let α be an o.s. tensor norm on ONORM.*

(a) *If α is completely right-projective on ONORM $\times$ OBAN then $\alpha = \alpha^{\to}$ on ONORM $\times$ OBAN.*

(b) *If $\alpha = \alpha^{\to}$ on ONORM $\times$ OBAN then $\alpha = \alpha^{\to}$ on ONORM $\times$ ONORM and for all $E, F \in$ ONORM with $\overline{F}$ the completion of F, the following canonical embedding is a complete isometry:*

$$E \otimes_\alpha F \hookrightarrow E \otimes_\alpha \overline{F}.$$

(c) *If α is completely right-projective on ONORM $\times$ OBAN then it is completely right-projective on ONORM $\times$ ONORM.*

Proof

(a) For $G \in$ OBAN, by the construction mentioned in Sect. 1.1.3 there is a complete metric surjection $q_G : Z_G \twoheadrightarrow G$. Since Z_G is an ℓ_1 sum of finite dimensional spaces it is clear that it has the completely metric approximation property. By the Approximation Lemma 4.1.1, for every normed operator space E we have that $\alpha = \alpha^{\to}$ on $E \otimes Z_G$.

Hence, for $z \in M_n(E \otimes G)$ and $\varepsilon > 0$ there exist $N \in$ OFIN(Z_G) and $u \in E \otimes N$ such that $Id_E \otimes q_G(u) = z$ and

$$\alpha_n(u; E, N) \leq (1 + \varepsilon)\alpha_n(z; E, G).$$

Therefore,

$$\alpha_n(z; E, G) \leq \alpha_n^{\to}(z; E, G) \leq \alpha_n(z; E, q_G(N)) \leq \alpha_n(u; E, N)$$
$$\leq (1 + \varepsilon)\alpha_n(z; E, G).$$

(b) Let $z \in M_n(E \otimes F)$. By the metric mapping property, $\alpha_n^{\rightarrow}(z; E, \overline{F}) \leq \alpha_n^{\rightarrow}(z; E, F)$. For $\varepsilon > 0$ let $N \in \mathrm{OFIN}(\overline{F})$ with $z \in M_n(E \otimes N)$ such that $\alpha_n(z; E, N) \leq (1 + \varepsilon)\alpha_n^{\rightarrow}(z; E, \overline{F})$. By Lemma 10.3.1 there is a completely bounded linear map $R : N \to F$ with $\|R\|_{\mathrm{cb}} \leq 1 + \varepsilon$ and $Ry = y$ for all $y \in F \cap N$. Note that $z \in M_n(E \otimes (F \cap N))$ because each entry of the matrix can be seen as an operator from E' to F and also from E' to N. Thus, we have $(Id_E \otimes R)_n(z) = z \in M_n(E \otimes R(N))$. Consequently,

$$\alpha_n^{\rightarrow}(z; E, F) \leq \alpha_n((Id_E \otimes R)_n(z); E, R(N)) \leq (1 + \varepsilon)\alpha_n(z; E, N)$$

$$\leq (1 + \varepsilon)^2 \alpha_n^{\rightarrow}(z; E, \overline{F}).$$

(c) Let $E, F, G \in \mathrm{ONORM}$ and $q : F \twoheadrightarrow G$ a complete metric surjection. By Lemma 10.3.2 the completion $\widetilde{q} : \overline{F} \twoheadrightarrow \overline{G}$ is also a complete metric surjection with $\ker \widetilde{q} = \overline{\ker q}$. We know from (a) and (b) that the following embeddings are complete isometries:

$$E \otimes_\alpha F \hookrightarrow E \otimes_\alpha \overline{F} \quad \text{and} \quad E \otimes_\alpha G \hookrightarrow E \otimes_\alpha \overline{G},$$

so, $Id_E \otimes \widetilde{q}(E \otimes_\alpha F) = E \otimes_\alpha G$. Now, our assumption is that $Id_E \otimes \widetilde{q} : E \otimes_\alpha \overline{F} \to E \otimes_\alpha \overline{G}$ is a complete metric surjection. Thus, applying again Lemma 10.3.2, in order to obtain that $Id_E \otimes q : E \otimes_\alpha F \to E \otimes_\alpha G$ is a complete metric surjection we just need to check that $\ker(Id_E \otimes \widetilde{q}) \subset \overline{\ker(Id_E \otimes q)}$ (where the closure is considered inside $E \otimes_\alpha \overline{F}$). Indeed, this is true because

$$\ker(Id_E \otimes \widetilde{q}) = E \otimes \ker \widetilde{q} = E \otimes \overline{\ker q} \subset \overline{E \otimes \ker q} = \overline{\ker(Id_E \otimes q)}.$$

$$\square$$

Remark 10.3.5 Observe that, using the previous lemma, if α is completely right-projective on ONORM then it is finitely-generated from the right, i.e., $\alpha = \alpha^{\rightarrow}$.

Theorem 10.3.6 *Let α be an o.s. tensor norm on* ONORM. *Then there is a unique completely right-projective o.s. tensor norm $\alpha/$ on* ONORM *such that $\beta \geq \alpha/$ for all completely right-projective o.s. tensor norms $\beta \geq \alpha$.*

For $E \in$ ONORM and $F \in$ OBAN, if Z_0 is a completely projective Banach operator space and $q : Z_0 \to F$ is a complete metric surjection, then

$$E \otimes_\alpha Z_0 \to E \otimes_{\alpha/} F$$

is a complete metric surjection.

We recall that, given a complete operator space E, the completely projective space Z_E introduced in Sect. 1.1.3 is locally reflexive (as mentioned at the end of Sect. 1.5.1). With this property at hand we can now prove Theorem 10.3.6.

Proof of Theorem 10.3.6 Let $F \in$ OBAN. Suppose that we have two complete quotients $q_j : Z_j \twoheadrightarrow F$, $j = 1, 2$ where the spaces Z_j are completely projective operator spaces, and suppose that we have operator space structures α_j on $E \otimes F$ making

$$id_E \otimes q_j : E \otimes_\alpha Z_j \to E \otimes_{\alpha_j} F$$

into complete quotients. Fix $\varepsilon > 0$. By the projectivity of Z_i, there exist operators $L_1 : Z_1 \to Z_2$ and $L_2 : Z_2 \to Z_1$ such that $\|L_i\|_{cb} \le 1 + \varepsilon$ satisfying $q_2 \circ L_1 = q_1$ and $q_1 \circ L_2 = q_2$. Observe that by the metric mapping property of α, $id_E \otimes L_1 : E \otimes_\alpha Z_1 \to E \otimes_\alpha Z_2$ and $id_E \otimes L_2 : E \otimes_\alpha Z_2 \to E \otimes_\alpha Z_1$ have cb-norm at most $1 + \varepsilon$. Since $id_E \otimes q_j : E \otimes_\alpha Z_j \to E \otimes_{\alpha_j} F$ is a complete quotient, it follows from [65, Prop. 2.4.1] that the identity mappings $E \otimes_{\alpha_1} F \to E \otimes_{\alpha_2} F$ and $E \otimes_{\alpha_2} F \to E \otimes_{\alpha_1} F$ have cb-norm at most $1 + \varepsilon$. Letting $\varepsilon \to 0$, we conclude that $E \otimes_{\alpha_1} F$ and $E \otimes_{\alpha_2} F$ are canonically completely isometric. Therefore, there is no ambiguity if we define $\alpha/$ on $E \otimes F$ to be the operator space structure on $E \otimes F$ induced by the particular quotient

$$E \otimes_\alpha Z_F \twoheadrightarrow E \otimes_{\alpha/} F$$

where Z_F is the projective space introduced in Sect. 1.1.3.

From the projectivity of proj, the fact that $\alpha \le$ proj and the diagram

$$
\begin{array}{ccc}
E \otimes_{\mathrm{proj}} Z_F & \longrightarrow\!\!\!\rightarrow & E \otimes_{\mathrm{proj}} F \\
\downarrow & & \vdots\!\!\downarrow \\
E \otimes_\alpha Z_F & \longrightarrow\!\!\!\rightarrow & E \otimes_{\alpha/} F
\end{array}
$$

together with [65, Prop. 2.4.1] it follows that $\alpha/ \le$ proj.

Similarly, from the inequality min $\le \alpha$ and the diagram

$$
\begin{array}{ccc}
E \otimes_\alpha Z_F & \longrightarrow\!\!\!\rightarrow & E \otimes_{\alpha/} F \\
\downarrow & & \vdots\!\!\downarrow \\
E \otimes_{\mathrm{min}} Z_F & \longrightarrow & E \otimes_{\mathrm{min}} F
\end{array}
$$

it follows that min $\le \alpha/$. Therefore, $\alpha/$ is a reasonable operator space cross-norm.

Let us now show that $\alpha/$ has the complete metric mapping property. Let $S \in$ CB(E_1, E_2) and $T \in$ CB(F_1, F_2) and consider complete quotients $Z_j \twoheadrightarrow F_j$ with the spaces Z_j being completely projective. Therefore, given $\varepsilon > 0$ there exists a completely bounded map $\widehat{T} : Z_1 \to Z_2$ such that $\|\widehat{T}\|_{cb} \le (1 + \varepsilon) \|T\|_{cb}$ and such that the diagram

$$
\begin{array}{ccc}
Z_1 & \xrightarrow{\;\widehat{T}\;} & Z_2 \\
\downarrow & & \downarrow \\
F_1 & \xrightarrow{\;T\;} & F_2
\end{array}
$$

commutes. After tensorizing, we obtain

$$
\begin{array}{ccc}
E_1 \otimes_\alpha Z_1 & \xrightarrow{\;S \otimes \widehat{T}\;} & E_2 \otimes_\alpha Z_2 \\
\downarrow & & \downarrow \\
E_1 \otimes_{\alpha/} F_1 & \xrightarrow{\;S \otimes T\;} & E_2 \otimes_{\alpha/} F_2
\end{array}
$$

from where we clearly see, appealing to the complete metric mapping property of α, that

$$
\left\| S \otimes T : E_1 \otimes_{\alpha/} F_1 \to E_2 \otimes_{\alpha/} F_2 \right\|_{\mathrm{cb}} \le (1 + \varepsilon) \, \|S\|_{\mathrm{cb}} \, \|T\|_{\mathrm{cb}} \, .
$$

Letting $\varepsilon \to 0$ yields the complete metric mapping property of $\alpha/$.

Now, let $G \in \mathbf{OBAN}$ and suppose that $q : F \twoheadrightarrow G$ is a complete quotient. If $q_F : Z_F \twoheadrightarrow F$ is the standard complete quotient, observe that the composition $q \circ q_F : Z_F \twoheadrightarrow G$ is also a complete quotient. For $z \in M_n(E \otimes G)$, by what was shown at the beginning of the proof we have

$$
\begin{aligned}
\alpha/_n(z; E, G) &= \inf \left\{ \alpha_n(w; E, Z_F) \; : \; (q \circ q_F)_n w = z \right\} \\
&= \inf \left\{ \alpha_n(w; E, Z_F) \; : \; (q_F)_n w = u,\, q_n u = z \right\} \\
&= \inf \left\{ \inf\{ \alpha_n(w; E, Z_F) \; : \; (q_F)_n w = u \} \; : \; q_n u = z \right\} \\
&= \inf \left\{ \alpha/_n(u; E, F) \; : \; q_n u = z \right\}.
\end{aligned}
$$

That is, $id_E \otimes q : E \otimes_{\alpha/} F \to E \otimes_{\alpha/} G$ is a complete quotient and therefore $\alpha/$ is completely right-projective in $\mathbf{ONORM} \times \mathbf{OBAN}$.

According to Lemma 10.3.4 (a) we know that the norm $\alpha/$ is finitely-generated from the right in $\mathbf{ONORM} \times \mathbf{OBAN}$. Thus, by Lemma 10.3.4 (b) and (c) this norm can be extended to $\mathbf{ONORM} \times \mathbf{ONORM}$ and this extension is completely right-projective on $\mathbf{ONORM}$. Moreover, by definition $\alpha \le \alpha/$ on $\mathbf{ONORM} \times \mathbf{OFIN}$, which implies that

$$
\alpha \le \alpha^{\rightarrow} \le (\alpha/)^{\rightarrow} = \alpha/.
$$

Finally, let β be any completely right-projective norm that dominates α. Then, $\alpha/ \le \beta$ in $\mathbf{ONORM} \times \mathbf{OFIN}$ by definition. Remark 10.3.5 says that $\beta = \beta^{\rightarrow}$ on

ONORM then, by Lemma 10.3.4 (a), we have

$$\alpha/ = (\alpha/)^{\rightarrow} \leq \beta^{\rightarrow} = \beta.$$

$\square$

Lemma 10.3.7 *If α be a finitely-generated o.s. tensor norm in* ONORM, *then $\alpha/$ is also finitely-generated.*

Proof Let $z \in M_n(E \otimes F)$. By Lemma 10.3.4 we know that $\alpha/$ is finitely-generated on the right. Then, for a given $\varepsilon > 0$ there is a subspace $N \in \mathrm{OFIN}(F)$ satisfying

$$(\alpha/)_n(z; E, N) \leq (1 + \varepsilon)(\alpha/)_n(z; E, F).$$

Since N is finite dimensional it is complete. Hence, there is a completely projective Banach operator space Z_N such that

$$id_E \otimes q_N : E \otimes_\alpha Z_N \twoheadrightarrow E \otimes_{\alpha/} N$$

is a complete quotient. Then there is an element $w \in M_n(E \otimes Z_N)$ with $(id_E \otimes q_N)_n(w) = z$ and

$$\alpha_n(w; E, Z_N) \leq (1 + \varepsilon)(\alpha/)_n(z; E, N).$$

Now, due to α being finitely-generated there exists $M \in \mathrm{OFIN}(E)$ such that

$$\alpha_n(w; M, Z_N) \leq (1 + \varepsilon)\alpha_n(w; E, Z_N).$$

Note that $id_M \otimes q_N : E \otimes_\alpha Z_N \to E \otimes_{\alpha/} N$ has norm one and $(id_M \otimes q_N)_n(w) = z$, therefore

$$(\alpha/)_n(z; M, N) \leq \alpha_n(w; M, Z_N) \leq (1 + \varepsilon)^3 \alpha_n(z; E, F).$$

$\square$

The following lemma will be useful for our purposes. The norm $\backslash\alpha$ is defined similarly as $\alpha/$ with the obvious changes of spaces.

Lemma 10.3.8 *Let α be an o.s. tensor norm finitely-generated from the right then $\backslash\alpha$ is finitely-generated.*

Proof Let $E, F \in \mathrm{ONORM}$ and $z \in M_n(E \otimes F)$. By the left version of Remark 10.3.5 we know that $\backslash\alpha$ is finitely-generated from the left. Then, given $\varepsilon > 0$ there exists $M \in \mathrm{OFIN}(E)$ such that $z \in M_n(M \otimes F)$ and

$$(\backslash\alpha)_n(z; M, F) \leq (1 + \varepsilon)(\backslash\alpha)_n(z; E, F),$$

Applying the left version of Theorem 10.3.6 (and using that $M \in \mathrm{OFIN} \subset \mathrm{OBAN}$) we have that there exists $u \in M_n(Z_M \otimes F)$ satisfying $(q_M \otimes id_F)_n(u) = z$ and

$$\alpha_n(u; Z_M, F) \leq (1 + \varepsilon)(\backslash \alpha)_n(z; M, F).$$

Since α is finitely-generated from the right, there is a space $N \in \mathrm{OFIN}(F)$ such that $u \in M_n(Z_M \otimes N)$ and

$$\alpha_n(u; Z_M, N) \leq (1 + \varepsilon)\alpha_n(u; Z_M, F).$$

The mapping

$$q_M \otimes id_N : Z_M \otimes_\alpha N \twoheadrightarrow M \otimes_{\backslash \alpha} N$$

satisfies $(q_M \otimes id_N)_n(u) = z$. This implies that $z \in M_n(M \otimes N)$ and

$$(\backslash \alpha)_n(z; M, N) \leq \alpha_n(u; Z_M, N),$$

which completes the proof. $\qquad \square$

Remark 10.3.9 All the previous results about completely right-injective and completely right-projective o.s. tensor norms have left versions with analogous proofs. Specifically, there are valid left statements of Remark 10.1.2, Proposition 10.1.3, Corollary 10.1.4, Theorem 10.2.1, Lemma 10.2.2. Also, after defining the left-finite hull of an o.s. tensor norm, there are left adaptations of Lemma 10.3.4, Remark 10.3.5, Theorem 10.3.6, Lemmas 10.3.7 and 10.3.8.

Definition 10.3.10 The o.s. tensor norm $\alpha/$ given in Theorem 10.3.6 is called the *completely right-projective hull* of α. One can define analogously the *completely left-projective hull* of α, denoted $\backslash \alpha$, and the *completely projective hull*

$$\backslash \alpha/ := (\backslash \alpha)/ = \backslash(\alpha/).$$

Note that there is no ambiguity in the previous equality since $\backslash(\alpha/)$ and $(\backslash \alpha)/$ are equal. Indeed, let $E, F \in OBAN$ and Z_E and Z_F completely projective operator spaces such that

$$q_E : Z_E \twoheadrightarrow E \quad \text{and} \quad q_F : Z_E \twoheadrightarrow F$$

are complete quotients then we known that both norms $(\backslash \alpha)/$ and $\backslash(\alpha/)$ coincide on $E \otimes F$ and can be computed through the complete quotient

$$q_E \otimes q_F : Z_E \otimes_\alpha Z_F \twoheadrightarrow E \otimes_{\backslash \alpha/} F.$$

Using Remark 10.3.5 we know that $\alpha/$ is finitely-generated from the right and therefore, by Lemma 10.3.8, the o.s. tensor norm $\backslash(\alpha/)$ is finitely-generated. The same argument can be used to prove that $(\backslash\alpha)/$ is finitely-generated. Since both norms $\backslash(\alpha/)$ and $(\backslash\alpha)/$ coincide on OBAN and they are finitely-generated, they must be equal on ONORM.

The operator space $S_1(H)$ is not completely projective [9], but with respect to completely right- (or left-) projective hulls of a finitely-generated o.s. tensor norm it behaves as if it was. The reason behind this is the fact that $S_1(H)$ is an $\mathcal{OS}_{1,1+}$ space (see Sect. 1.1.1): each finite-dimensional subspace of $S_1(H)$ is contained in a larger finite-dimensional subspace which is "almost a copy" of S_1^k for some k:

Proposition 10.3.11 ([33, Prop. 5.3]) *Given $M \in \mathrm{OFIN}(S_1(H))$ and $\varepsilon > 0$ there exist a subspace $\widetilde{M} \in \mathrm{OFIN}(S_1(H))$, a number $k \in \mathbb{N}$ and a complete isomorphism $T : \widetilde{M} \to S_1^k$ such that $\widetilde{M} \supset M$, $\|T\|_{\mathrm{cb}} \leq 1 + \varepsilon$ and $\|T^{-1}\|_{\mathrm{cb}} \leq 1 + \varepsilon$.*

Now we can state and prove the result mentioned above.

Proposition 10.3.12 *Let α be a finitely-generated o.s. tensor norm and let $E \in$ ONORM. Then,*

$$E \otimes_\alpha S_1(H) = E \otimes_{\alpha/} S_1(H) \quad \text{and} \quad S_1(H) \otimes_\alpha E = S_1(H) \otimes_{\backslash\alpha} E.$$

Proof Since $\alpha \leq \alpha/$ in order to prove the first identity (the other is obtained by transposition) we have just to show for any $z \in M_n(E \otimes S_1(H))$ that $(\alpha/)_n(z; E, S_1(H)) \leq \alpha_n(z; E, S_1(H))$. Given $\varepsilon > 0$ there is $M \in \mathrm{OFIN}(S_1(H))$ satisfying $z \in M_n(E \otimes M)$ and

$$\alpha_n(z; E, M) \leq (1 + \varepsilon)\alpha_n(z; E, S_1(H)). \tag{10.3.1}$$

By Proposition 10.3.11 there are a finite-dimensional space $\widetilde{M} \supset M$, a number $k \in \mathbb{N}$ and a complete isomorphism $T : \widetilde{M} \to S_1^k$ such that $\|T\|_{\mathrm{cb}} \leq 1 + \varepsilon$ and $\|T^{-1}\|_{\mathrm{cb}} \leq 1 + \varepsilon$. Then

$$\alpha_n((id \otimes T)_n(z); E, S_1^k) \leq (1+\varepsilon)\alpha_n(z; E, \widetilde{M}) \leq (1+\varepsilon)\alpha_n(z; E, M). \tag{10.3.2}$$

Now, applying T^{-1} and using that S_1^k is completely projective we have

$$(\alpha/)_n(z; E, \widetilde{M}) \leq (1 + \varepsilon)\alpha_n((id \otimes T)_n(z); E, S_1^k). \tag{10.3.3}$$

Finally, since $(\alpha/)_n(z; E, S_1(H)) \leq (\alpha/)_n(z; E, \widetilde{M})$ compounding the identities (10.3.1), (10.3.2) and (10.3.3) we arrive to

$$(\alpha/)_n(z; E, S_1(H)) \leq (1 + \varepsilon)^3 \alpha_n(z; E, S_1(H)),$$

which concludes the proof. $\square$

As a consequence of the previous result we present the following operator space version of a classical Banach space identity.

Example 10.3.13 $d_\infty = \min /$.

Indeed, since both norms are finitely-generated (by Proposition 3.1.5 and Lemma 10.3.8) it is enough to prove that $M \otimes_{d_\infty} N = M \otimes_{\min /} N$ for every $M, N \in \mathrm{OFIN}$. Using that every separable operator space is completely isometric to a quotient of S_1 [65, Cor. 2.12.3] and that both d_∞ and $\min /$ are completely right-projective, we know that there is a complete quotient $q : S_1 \twoheadrightarrow N$ which produces two other complete quotients

$$id \otimes q : M \otimes_{d_\infty} S_1 \twoheadrightarrow M \otimes_{d_\infty} N \quad \text{and} \quad id \otimes q : M \otimes_{\min /} S_1 \twoheadrightarrow M \otimes_{\min /} N.$$

Hence, the result is obtained once we prove $M \otimes_{d_\infty} S_1 = M \otimes_{\min /} S_1$ for every $M \in \mathrm{OFIN}$. This is certainly true because $M \otimes_{d_\infty} S_1 = M \otimes_{\min} S_1$ by Theorem 2.2.8 and $M \otimes_{\min /} S_1 = M \otimes_{\min} S_1$ (by Proposition 10.3.12).

Another consequence of Propositions 10.3.11 and 10.3.12 is the following statement that will be useful later. The proof is obtained proceeding as in Example 10.3.13 along with Proposition 10.3.12.

Corollary 10.3.14 *Let $E, F \in \mathrm{ONORM}$ with F separable and let α be a finitely-generated o.s. tensor norm. If $z \in M_n(E \otimes_\alpha / F)$ and $\varepsilon > 0$ there exist a number k, a mapping $R : S_1^k \to F$ and a matrix $u \in M_n(E \otimes_\alpha S_1^k)$ such that $(id \otimes R)_n(u) = z$, $\|R\|_{\mathrm{cb}} \le 1 + \varepsilon$ and*

$$\alpha_n(u; E, S_1^k) \le (1 + \varepsilon)(\alpha /)_n(z; E, F).$$

10.4 Local Description of Injective/Projective Hulls

A key result in the Banach space framework offers a straightforward criterion for determining whether a tensor norm β is the projective or injective associate of a given tensor norm α. Specifically:

Proposition 10.4.1 *[23, Prop. 20.9] Let α and β be tensor norms.*

(1) If β is right-projective, then the following statements are equivalent:

   (a) $\beta = \alpha /$
   (b) For every normed space E and every $n \in \mathbb{N}$

$$E \otimes_\beta \ell_1^n = E \otimes_\alpha \ell_1^n \quad \textit{isometrically.}$$

(2) If β is right-injective, then the following statements are equivalent:

   (a) $\beta = \alpha \backslash$
   (b) For every normed space E and every $n \in \mathbb{N}$

$$E \otimes_\beta \ell_\infty^n = E \otimes_\alpha \ell_\infty^n \quad \textit{isometrically.}$$

The statement in part (1) of the previous proposition extends canonically to the non-commutative context (with S_1^n in the place of ℓ_1^n), see Proposition 10.4.2 below. However, an analogous result in full generality for part (2) (clearly, with M_n in the place of ℓ_∞^n) is unattainable, as shown in Example 10.4.3. Indeed, this example explicitly provides two o.s. tensor norms, α and β, with β being completely right-injective, such that for every operator space E and every $n \in \mathbb{N}$, the equality $E \otimes_\beta M_n = E \otimes_\alpha M_n$ holds completely isometrically, although $\beta \neq \alpha\backslash$. While for a Banach space specialist at first glance it may seem counterintuitive that the theory cannot be fully transferred, the reason behind this difference is not a complete surprise from the operator space point of view: exactness. Indeed, the issue arises because the proof of part (2) relies on the fact that, in the Banach space setting, the injective hull is realized by embedding into an $\ell_\infty(I)$ space, which is an $\mathcal{L}_{\infty,1+}$ space (i.e., for every $\varepsilon > 0$, any finite-dimensional subspace of $\ell_\infty(I)$ is contained in a larger subspace that is $1 + \varepsilon$ isomorphic to an ℓ_∞^n). This allows the use of the $\mathcal{L}_p$-local Technique Lemma. In contrast, the argument does not carry over into the operator space setting. While we have a local Technique Lemma (Lemma 4.5.1), it only applies to $\mathcal{OS}_{p,C}$ spaces (see definition in Sect. 1.1.1). Notably, $\mathcal{OS}_{\infty,C}$ spaces are clearly exact (see definition in Sect. 1.5.2). Since in the operator space setting, the completely injective hull is given by embedding into a $\mathcal{B}(H)$ space, and $\mathcal{B}(H)$ is not exact when H is infinite-dimensional, the Banach space proof cannot be adapted to the operator space setting.

Nevertheless, as expected, a more modest result analogous to Proposition 10.4.1 (2) can be obtained within our framework if we restrict its applicability to the class of $\mathcal{OS}_{\infty,C}$ spaces (see Proposition 10.4.4, below).

Let us begin with the non-commutative counterpart of Proposition 10.4.1 (1).

Proposition 10.4.2 *Let α and β be o.s. tensor norms with β completely right-projective. The following are equivalent:*

(a) $\beta = \alpha/$.
(b) For every operator space E and every $n \in \mathbb{N}$, $E \otimes_\beta S_1^n = E \otimes_\alpha S_1^n$ completely isometrically.

Proof That (a) implies (b) is clear from Theorem 10.3.6 and the fact that S_1^n is completely projective.

Now, suppose that (b) holds. Since both o.s. tensor norms β and $\alpha/$ are completely right-projective, by Remark 10.3.5 they are finitely generated from the right. So, in order to prove (a) it suffices to show that these o.s. tensor norms coincide on $E \otimes F$ for any $E \in \mathrm{ONORM}$ and $F \in \mathrm{OFIN}$. To this end, consider a complete quotient $q_F : S_1 \twoheadrightarrow F$. The induced mappings

$$id_E \otimes_{\alpha/} q_F : E \otimes_{\alpha/} S_1 \longrightarrow E \otimes_{\alpha/} F \quad \text{and} \quad id_E \otimes_\beta q_F : E \otimes_\beta S_1 \longrightarrow E \otimes_\beta F$$

are complete quotients by the right-projectivity of the norms.

Now, if we establish that $E \otimes_\beta S_1 = E \otimes_{\alpha/} S_1$ completely isometrically, we would conclude that $\beta = \alpha/$ on $E \otimes F$ finishing the proof. For that, given the hypothesis and the fact that S_1^n is completely projective, together with Theorem 10.3.6, we obtain that for every n,

$$E \otimes_\beta S_1^n = E \otimes_{\alpha/} S_1^n$$

completely isometrically. The assertion then follows from Lemma 4.5.1 (and the last sentence of Remark 4.5.2) since S_1 is an $\mathscr{OS}_{1,1+}$ space, as mentioned in Proposition 10.3.11.

$\square$

The following example shows that we cannot expect to have an analogous version of Proposition 10.4.1 (2) in the operator space setting.

Example 10.4.3 There exist two o.s. tensor norms, α and β, with β being completely right-injective, such that for every operator space E and every $n \in \mathbb{N}$, the equality $E \otimes_\beta M_n = E \otimes_\alpha M_n$ holds completely isometrically, although $\beta \neq \alpha\backslash$.

Proof Let $\alpha := \backslash\min$ and $\beta := (\alpha'/)'$. First, observe that by Corollary 10.1.4, β is completely right-injective. Additionally, both o.s. tensor norms are finitely-generated: for α, this follows from Lemma 10.3.8, and for β, this is straightforward since it is a dual tensor norm.

To verify that for every operator space E and every $n \in \mathbb{N}$, the equality $E \otimes_\beta M_n = E \otimes_\alpha M_n$ holds completely isometrically, it suffices to check it for every $E \in$ OFIN and every $n \in \mathbb{N}$. Indeed, for any fixed finite-dimensional operator space E and any fixed n, the following complete isometries hold:

$$E \otimes_\beta M_n = E \otimes_{(\alpha'/)'} M_n = \left(E' \otimes_{\alpha'/} S_1^n\right)' = \left(E' \otimes_{\alpha'} S_1^n\right)' = E \otimes_\alpha M_n.$$

Nevertheless, $\beta \neq \alpha\backslash$. To see this, first note that $\alpha'/$ is finitely-generated by Lemma 10.3.7 and is therefore equal to β'. If β were equal to $\alpha\backslash$, then by taking duals, $\beta' = \alpha'/$ would coincide with $(\alpha\backslash)'$. However, this is not the case, as shown in Remark 12.1.5 (b) below.

$\square$

We now consider a (more limited) version of Proposition 10.4.1 (2) within our framework.

Proposition 10.4.4 *Let α and β be operator space tensor norms, with β completely right-injective. If $E \otimes_\beta M_n = E \otimes_\alpha M_n$ for all $n \in \mathbb{N}$, then for any $\mathscr{OS}_{\infty,C}$ space F, the spaces $E \otimes_\beta F$ and $E \otimes_{\alpha\backslash} F$ are completely isomorphic. Namely, $C^{-1}\beta \leq \alpha\backslash \leq C\beta$ on $E \otimes F$.*

Proof Since both o.s. tensor norms β and $\alpha\backslash$ are completely right-injective, they are obviously also right-finitely-generated (i.e., they coincide with their right-finite hulls). Therefore, the result follows from Lemma 4.5.1 taking into account the comments of Remark 4.5.2.

$\square$

Chapter 11
Injective/Projective Hulls and Accessibility

In Corollary 10.1.4 we have shown that the dual of a completely right-projective (completely left-projective) o.s. tensor norm is completely right-injective (completely left-injective). In the Banach space setting the reciprocal is valid but in the operator space context this is no longer true (see Remark 12.1.5 below). Note that all known proofs of this fact for Banach space tensor norms use elements of the theory mentioned in Sect. 1.5.

Hence, in order to produce a result relating completely injective/projective hulls with duality we have to appeal for an extra hypothesis of accessibility.

11.1 The Impact of Accessibility

Even if previously we presented in detail results about right hulls and obtain by analogy the left statements, now, *to see the other side of the moon,* we chose to proceed first with left hulls and to derive by similarity the right versions.

Proposition 11.1.1 *Let α be a finitely-generated o.s. tensor norm.*

(a) If α is left-accessible then $(\backslash\alpha')' = /\alpha$ and $(/\alpha)' = \backslash\alpha'$.
(b) If α is right-accessible then $(\alpha'/)' = \alpha\backslash$ and $(\alpha\backslash)' = \alpha'/$.
(c) If α is totally accessible then $(\backslash\alpha'/)' = /\alpha\backslash$ and $(/\alpha\backslash)' = \backslash\alpha'/$.

Proof We will only prove (a) since (b) and (c) follow analogously. To see the first identity, it suffices to check it in OFIN since both norms are finitely-generated. Indeed, the norm $(\backslash\alpha')'$ is finitely-generated since it is a dual norm and $/\alpha$ is so by the left version of Lemma 10.2.2.

Let M and N be finite dimensional operator spaces. We have to see that

$$\left(M \otimes_{\backslash\alpha'} N'\right)' = M' \otimes_{/\alpha} N. \tag{11.1.1}$$

© The Author(s), under exclusive license to Springer Nature Switzerland AG 2025

J. A. Chávez-Domínguez et al., *Operator Space Tensor Norms*, Lecture Notes in Mathematics 2379, https://doi.org/10.1007/978-3-031-96209-7_11

To check (11.1.1) we consider the following diagram

$$
\begin{array}{ccc}
(M \otimes_{\backslash\alpha'} N')' & \xrightarrow{\quad\quad\quad (A) \quad\quad\quad} & (Z_M \otimes_{\backslash\alpha'} N')' \ , \\
{\scriptstyle(\spadesuit)}\Big\downarrow & & \Big\| {\scriptstyle(B)} \\
M' \otimes_{/\alpha} N \ \hookrightarrow\ Z'_M \otimes_{/\alpha} N \stackrel{(C)}{=} Z'_M \otimes_\alpha N & \xhookrightarrow{\ (D)\ } & (Z_M \otimes_{\alpha'} N')'
\end{array}
$$

Note that the arrow (A) is completely isometric because it is the transpose of the complete quotient given by

$$
Z_M \otimes_{\backslash\alpha'} N' \ \longrightarrow\!\!\!\rightarrow\ M \otimes_{\backslash\alpha'} N'.
$$

Equalities in (B) and (C) hold since Z_M is completely projective and Z'_M is completely injective [65, Cor. 24.6], respectively. The mapping (D) is a complete isometry by the Duality Theorem 5.2.1 (recall that Z_M is locally reflexive) and the equality $\alpha = \overleftarrow{\alpha}$ in $Z'_M \otimes N$ which holds since α is left-accessible. Therefore, the mapping $(\spadesuit)$ is a completely isometric isomorphism.

The second identity of (a) follows by taking adjoints and using that, by the left version of Lemma 10.3.7, $\backslash\alpha'$ is finitely-generated. $\qquad\square$

Remark 11.1.2 Looking at the bottom line of the previous commutative diagram, we have that if α is a left-accessible finitely-generated o.s. tensor norm and M and N are finite dimensional operator spaces then we have the following complete isometry:

$$
M' \otimes_{/\alpha} N \hookrightarrow (Z_M \otimes_{\alpha'} N')'.
$$

Similarly, we have for α right-accessible and finitely-generated:

$$
M' \otimes_{\alpha\backslash} N \hookrightarrow (M \otimes_{\alpha'} Z_{N'})'.
$$

Example 11.1.3 Since proj is left- and right-accessible and finitely-generated, Proposition 11.1.1 tells us that

$$
(\backslash\min)' = /\operatorname{proj} \quad \text{and} \quad (\min/)' = \operatorname{proj}\backslash.
$$

It would be interesting to know whether the equality $(\backslash\min/)' = /\operatorname{proj}\backslash$ holds. This can not be easily derived from Proposition 11.1.1 (a) since we do not know if $\operatorname{proj}\backslash$ is left-accessible, or from Proposition 11.1.1 (c) since we know that proj is not totally accessible. Later, in Remark 12.4.3 we will see that these two norms, $(\backslash\min/)'$ and $/\operatorname{proj}\backslash$ are equivalent at the Banach space level. The question about if they are equal (or, at least, equivalent) as o.s. tensor norms remains open.

11.2 Accessibility and Mapping Procedures

As we have seen, accessibility is related with the duality between injective and projective hulls of o.s. tensor norms. This property also appears when describing the associated norms of surjective and injective hulls of mapping ideals (see Theorem 11.2.3 below). First, we need the following lemma.

Lemma 11.2.1 *Let* $(\mathfrak{A}, \mathbf{A})$ *be a mapping ideal.*

(a) If β is a finitely-generated o.s. tensor norm associated to $\mathfrak{A}^{\mathrm{sur}}$ then β is completely left-injective.

(b) If β is a finitely-generated o.s. tensor norm associated to $\mathfrak{A}^{\mathrm{inj}}$ then β is completely right-injective.

Proof

(a) We suppose first that E_1, E_2, F are finite dimensional and $i : E_1 \to E_2$ is a complete isometry. Then,

$$E_1 \otimes_\beta F = \mathfrak{A}^{\mathrm{sur}}(E_1', F)$$

$$E_2 \otimes_\beta F = \mathfrak{A}^{\mathrm{sur}}(E_2', F).$$

Through these identifications the mapping $i \otimes id_F : E_1 \otimes_\beta F \to E_2 \otimes_\beta F$ looks like

$$\mathfrak{A}^{\mathrm{sur}}(E_1', F) \longrightarrow \mathfrak{A}^{\mathrm{sur}}(E_2', F)$$

$$T \longmapsto T \circ i'.$$

Since $\mathfrak{A}^{\mathrm{sur}}$ is surjective and $i' : E_2' \twoheadrightarrow E_1'$ is a complete quotient mapping, we have $\mathbf{A}_n^{\mathrm{sur}}(T) = \mathbf{A}_n^{\mathrm{sur}}(T \circ i')$ for all $T \in M_n(\mathfrak{A}^{\mathrm{sur}}(E_1', F))$, which concludes the proof in the finite dimensional case.

Now, the conclusion follows due to the left version of Proposition 10.1.3.

(b) As in the previous item, we begin by assuming that E, F_1, F_2 are finite dimensional operator spaces and $i : F_1 \to F_2$ is a complete isometry. Then,

$$E \otimes_\beta F_1 = \mathfrak{A}^{\mathrm{inj}}(E', F_1)$$

$$E \otimes_\beta F_2 = \mathfrak{A}^{\mathrm{inj}}(E', F_2).$$

Through these identifications the mapping $id_E \otimes i : E \otimes_\beta F_1 \to E \otimes_\beta F_2$ looks like

$$\mathfrak{A}^{\mathrm{inj}}(E', F_1) \longrightarrow \mathfrak{A}^{\mathrm{inj}}(E', F_2)$$

$$T \longmapsto i \circ T.$$

Now, the equality $\mathbf{A}_n^{\mathrm{inj}}(T) = \mathbf{A}_n^{\mathrm{inj}}(i_n \circ T)$ for all $T \in M_n(\mathfrak{A}^{\mathrm{inj}}(E', F_1))$ concludes the proof in the finite dimensional case. As in the previous item the case for arbitrary operator spaces follows from Proposition 10.1.3.

$\square$

Remark 11.2.2 Let α, β, γ be finitely-generated o.s. tensor norms associated to the mapping ideals $\mathfrak{A}$, $\mathfrak{A}^{\mathrm{sur}}$ and $\mathfrak{A}^{\mathrm{inj}}$ respectively. By the previous lemma we have

$$\beta \leq /\alpha \quad \text{and} \quad \gamma \leq \alpha \backslash. \tag{11.2.1}$$

Theorem 11.2.3 *Let* $(\mathfrak{A}, \mathbf{A})$ *be a mapping ideal and* α *its associated finitely-generated o.s. tensor norm.*

(a) If α *is left-accessible then the o.s. tensor norm* $/\alpha$ *is associated to* $\mathfrak{A}^{\mathrm{sur}}$.
(b) If α *is right-accessible then the o.s. tensor norm* $\alpha \backslash$ *is associated to* $\mathfrak{A}^{\mathrm{inj}}$.

Proof

(a) Suppose first that $\mathfrak{A}$ is a maximal mapping ideal and let M, N be finite-dimensional operator spaces. Let $q_M : Z_M \twoheadrightarrow M$ be the canonical complete quotient mapping. Then, for every $T \in M_n(\mathfrak{A}^{\mathrm{sur}}(M, N))$ we have

$$\mathbf{A}_n^{\mathrm{sur}}(T) = \mathbf{A}_n(T \circ q_M) = \|T \circ q_M\|_{M_n((Z_M \otimes_{\alpha'} N')')} = (/\alpha)_n(T; M', N),$$

where the second equality follows from the Representation Theorem 8.2.3 and the third comes from Remark 11.1.2. Therefore, we have the complete isometry

$$\mathfrak{A}^{\mathrm{sur}}(M, N) \overset{1}{=} M' \otimes_{/\alpha} N.$$

In particular, $/\alpha$ is the associate o.s. tensor norm of $\mathfrak{A}^{\mathrm{sur}}$.

Now, let $\mathfrak{A}$ be an arbitrary mapping ideal. Note that if $\mathfrak{A} \sim \alpha$ then $\mathfrak{A}^{\mathrm{max}} \sim \alpha$ and thus, by the first part of the proof $(\mathfrak{A}^{\mathrm{max}})^{\mathrm{sur}} \sim /\alpha$.

The inclusion $\mathfrak{A} \subset \mathfrak{A}^{\mathrm{max}}$ (in the o.s. setting) implies $\mathfrak{A}^{\mathrm{sur}} \subset (\mathfrak{A}^{\mathrm{max}})^{\mathrm{sur}}$. Denoting by β the finitely-generated o.s. tensor norm associated to $\mathfrak{A}^{\mathrm{sur}}$ we derive that

$$/\alpha \leq \beta. \tag{11.2.2}$$

Now the equality follows by Remark 11.2.2.

(b) Suppose first that $\mathfrak{A}$ is a maximal mapping ideal and let M, N be finite-dimensional operator spaces. Let $q_{N'} : Z_{N'} \twoheadrightarrow N'$ be the canonical complete quotient mapping. Then, for every $T \in M_n(\mathfrak{A}^{\mathrm{inj}}(M, N))$ we have

$$\mathbf{A}_n^{\mathrm{inj}}(T) = \mathbf{A}_n((q'_{N'})_n \circ T) = \|(q'_{N'})_n \circ T\|_{M_n((M \otimes_{\alpha'} Z_{N'})')} = (\alpha \backslash)_n(T; M', N),$$

where the second equality follows from the Representation Theorem 8.2.3 and the third comes from Remark 11.1.2. Therefore, we have the complete isometry

$$\mathfrak{A}^{\mathrm{inj}}(M, N) \overset{1}{=} M' \otimes_{\alpha\backslash} N.$$

In particular, $\alpha\backslash$ is an o.s. tensor norm associated to $\mathfrak{A}^{\mathrm{inj}}$.

The proof for an arbitrary mapping ideal $\mathfrak{A}$ runs as in the previous item.

$\square$

11.3 Accessibility of Mapping Ideals

Now we study the relation between accessibility of a tensor norm with some properties of its associated mapping ideal. As it is our habit in the operator space framework, we consider local reflexivity versions of the definitions.

Definition 11.3.1 A mapping ideal $(\mathfrak{A}, \mathbf{A})$ is called *right-accessible (locally right-accessible)* if for all $M \in \mathrm{OFIN}$, $F \in \mathrm{OBAN}$ ($F \in \mathrm{OLOC}$), $n \in \mathbb{N}$, $T \in M_n(\mathrm{CB}(M, F))$ and $\varepsilon > 0$ there are $N \in \mathrm{OFIN}(F)$ and $S \in M_n(\mathrm{CB}(M, N))$ such that $\mathbf{A}_n(S) \leq (1 + \varepsilon)\mathbf{A}_n(T)$ and the following diagram commutes

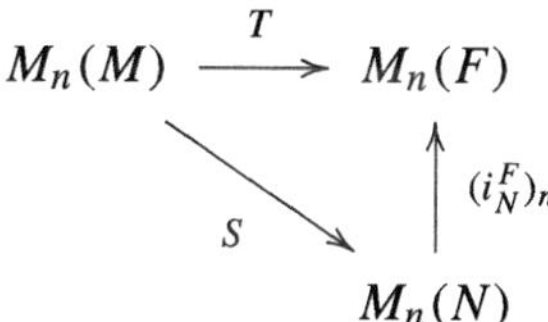

$(\mathfrak{A}, \mathbf{A})$ is called *left-accessible (locally left-accessible)* if for all $E \in \mathrm{OBAN}$ ($E \in \mathrm{OLOC}$), $N \in \mathrm{OFIN}$, $n \in \mathbb{N}$, $T \in M_n(\mathrm{CB}(E, N))$ and $\varepsilon > 0$, there exist $L \in \mathrm{OCOFIN}(E)$ and $S \in M_n(\mathrm{CB}(E/L, N))$ such that $\mathbf{A}_n(S) \leq (1 + \varepsilon)\mathbf{A}_n(T)$ and the following diagram commutes

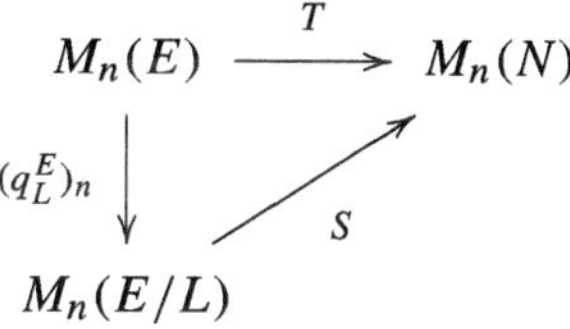

A mapping ideal which is both left and right-accessible (locally left- and locally right-accessible) is called *accessible (locally accessible)*. Moreover, $(\mathfrak{A}, \mathbf{A})$ is called *totally accessible (locally totally accessible)* if for every matrix of finite-rank operators $T : M_n(E) \to M_n(F)$ between complete operator spaces (complete locally reflexive operator spaces) and $\varepsilon > 0$, there are $L \subset \mathrm{OCOFIN}(E)$, $N \subset \mathrm{OFIN}(F)$ and $S \in M_n(\mathrm{CB}(E/L, N))$ such that $\mathbf{A}_n(S) \leq (1 + \varepsilon)\mathbf{A}_n(T)$ and the following diagram commutes

$$
\begin{array}{ccc}
M_n(E) & \xrightarrow{\ \ T\ \ } & M_n(F) \\
\big\downarrow{\scriptstyle (q_L^E)_n} & & \big\uparrow{\scriptstyle (i_N^F)_n} \\
M_n(E/L) & \xrightarrow[\ \ S\ \ ]{} & M_n(N)
\end{array}
$$

The necessity of these *local* versions of the accessibility definitions is justified by Example 11.3.6 (2), where we show that the mapping ideal $\mathcal{I}$ is locally left-accessible but not left-accessible.

Remark 11.3.2 It is clear that every completely injective mapping ideal is right-accessible, and every completely projective mapping ideal is left-accessible. Assume $T : M_n(E) \to M_n(F)$ is a matrix of finite-rank operators. Let $\mathrm{im}(T)$ denote the span of the union of the spaces $(T_{ij}(E))_{i,j=1}^n$, and observe that $\mathrm{im}(T) \in$ OFIN(F). Denoting $\ker(T) = \bigcap_{i,j=1}^n \ker(T_{ij})$, note $\ker(T) \in$ OCOFIN(E) and moreover there is a canonical factorization

$$
\begin{array}{ccc}
M_n(E) & \xrightarrow{\ \ T\ \ } & M_n(F) \\
\big\downarrow & & \big\uparrow \\
M_n(E/\ker(T)) & \xrightarrow[\ \ S\ \ ]{} & M_n(\mathrm{im}(T))
\end{array}
$$

This shows that a mapping ideal that is both completely injective and completely projective is totally accessible.

As an example, since we have already noted that CB is both completely injective and completely projective, we deduce that the mapping ideal CB is totally accessible.

Remark 11.3.3 Note that if $(\mathfrak{A}^{\mathrm{max}}, \mathbf{A}^{\mathrm{max}})$ is right-accessible (left-accessible, totally accessible, locally accessible) then $(\mathfrak{A}, \mathbf{A})$ is right-accessible (left-accessible, totally accessible, locally accessible). Indeed, this is clear from the definitions of the different variants of accessibility along with the completely contractive inclusion $\mathfrak{A}(E, F) \subset \mathfrak{A}^{\mathrm{max}}(E, F)$ for $E, F \in$ OBAN and the completely isometric equality $\mathfrak{A}(M, N) = \mathfrak{A}^{\mathrm{max}}(M, N)$ for $M, N \in$ OFIN.

There is a reason why we are defining notions of accessibility for both cross-norms and mapping ideals: in the Banach space setting, for a maximal mapping ideal $\mathfrak{A}$ associated with a (finitely-generated) cross-norm α, the two notions of accessibility coincide. We will now show that, in the operator space world, there is a relationship between (some of) the notions of accessibility for o.s. tensor norms and for operator ideals, but, as usual, local reflexivity is involved. Actually, for right accessibility the equivalence is just as in the Banach space case but for left accessibility we could only prove a weaker statement.

Proposition 11.3.4 *Let $(\mathfrak{A}, \mathbf{A})$ be a mapping ideal and α be its associated finitely-generated o.s. tensor norm. Then:*

(a) α is right-accessible (locally right-accessible) if and only if $(\mathfrak{A}, \mathbf{A})$ is right-accessible (locally right-accessible).

(b) If α is left-accessible (totally accessible) then $(\mathfrak{A}, \mathbf{A})$ is locally left-accessible (locally totally accessible).

Proof

(a) Assume that α is right-accessible. Due to Remark 11.3.3 we may consider that $(\mathfrak{A}, \mathbf{A})$ is maximal. Let $M \in \mathrm{OFIN}$, $F \in \mathrm{OBAN}$ and $T \in M_n(\mathrm{CB}(M, F))$. Denote by $z_T \in M_n(M' \otimes F)$ the matrix of tensors corresponding to T. By the definition of right-accessibility and the Embedding Theorem 8.2.15 (using the fact that M is finite-dimensional and therefore locally reflexive) we have

$$\overrightarrow{\alpha}_n(z_T; M', F) = \overleftarrow{\alpha}_n(z_T; M', F) = \mathbf{A}_n(T). \qquad (11.3.1)$$

This implies that there are $\widetilde{M} \in \mathrm{OFIN}(M')$, $N \in \mathrm{OFIN}(F)$ and $u \in M_n(\widetilde{M} \otimes N)$ such that

$$\alpha_n(u; \widetilde{M}, N) \leq (1 + \varepsilon)\mathbf{A}_n(T) \quad \text{and} \quad (i_{\widetilde{M}}^{E'} \otimes i_N^F)_n(u) = z_T;$$

moreover, note that we can assume $\widetilde{M} = M'$. Hence, if $T_u \in M_n(\mathrm{CB}(M, N))$ is the matrix of mappings corresponding to u, it satisfies $\mathbf{A}_n(T_u) \leq (1 + \varepsilon)\mathbf{A}_n(T)$ and $(i_N^F)_n \circ T_u = T$.

Conversely, assume that $(\mathfrak{A}, \mathbf{A})$ is right-accessible. Since α is finitely-generated, $\alpha = \overrightarrow{\alpha}$. We always have $\overleftarrow{\alpha} \leq \alpha$, so all we need to show is

$$\alpha(\cdot; M', F) \leq \overleftarrow{\alpha}(\cdot; M', F)$$

for finite-dimensional M and arbitrary F. Given $z \in M_n(M' \otimes F)$ and $\varepsilon > 0$, by the right-accessibility of $\mathfrak{A}$ there are $N \in \mathrm{OFIN}(F)$ and $S \in M_n(\mathrm{CB}(M, N))$ such that

$$\mathbf{A}_n(S) \leq (1 + \varepsilon)\mathbf{A}_n(T_z) \quad \text{and} \quad (i_N^F)_n \circ S = T_z$$

where $T_z \in M_n(\mathrm{CB}(M, F))$ is the matrix of mappings corresponding to z. If $z_S \in M_n(E' \otimes N)$ is the matrix of tensors corresponding to S, since $M' \otimes_\alpha N = \mathfrak{A}(M, N)$ completely isometrically, it follows that

$$\alpha_n(z_S; M', N) = \mathbf{A}_n(S) \leq (1 + \varepsilon)\overleftarrow{\alpha}_n(z; E', F)$$

and $(i_{M'} \otimes i_N^F)_n(z_S) = z$, showing that α is right accessible.

The case of local right-accessibility follows in the same way just replacing $F \in \mathrm{OBAN}$ by $F \in \mathrm{OLOC}$.

(b) We prove the case of left-accessibility; the other one is similar. Again, by Remark 11.3.3 we may consider that $(\mathfrak{A}, \mathbf{A})$ is maximal. Let $E \in \mathrm{OLOC}$, $N \in \mathrm{OFIN}$ and $T \in M_n(\mathrm{CB}(E, N))$. Denoting by $z_T \in M_n(E' \otimes N)$ the matrix of tensors corresponding to T and applying the definition of left-accessibility and the Embedding Theorem 8.2.15 we have

$$\overrightarrow{\alpha}_n(z_T; E', N) = \overleftarrow{\alpha}_n(z_T; E', N) = \mathbf{A}_n(T).$$

This implies that there is $M \in \mathrm{OFIN}(E')$ and $u \in M_n(M \otimes N)$ such that

$$\alpha_n(u; M, N) \leq (1 + \varepsilon)\mathbf{A}_n(T) \quad \text{and} \quad (i_M^{E'} \otimes id)_n(u) = z_T.$$

Hence, if $T_u \in M_n(\mathrm{CB}(E/^0 M, N))$ is the matrix of mappings corresponding to u, we arrive to $\mathbf{A}_n(T_u) \leq (1 + \varepsilon)\mathbf{A}_n(T)$ and $T_u \circ q_{0_M}^E = T$, which finishes the proof.

$\square$

Remark 11.3.5 If α' is an $\mathcal{E}(\lambda)$-o.s. tensor norm, then the conclusion of Proposition 11.3.4 (b) is obtained without the word "locally". This is clear following the previous argument since, as we comment in Remark 8.2.16, in this case the Embedding Theorem 8.2.15 is valid without the local reflexivity hypothesis. On the other hand, in general it is not possible to obtain the conclusion of Proposition 11.3.4 (b) without the word "locally". Indeed, in Example 11.3.6 (2) we see that $\mathcal{I}$ is not left-accessible even though the associated o.s. tensor norm proj is left-accessible.

Example 11.3.6 Applying Proposition 11.3.4 and Remark 11.3.5 we derive the following examples:

(1) The mapping ideal $\mathfrak{A}_h$ is totally accessible.

Since h is totally accessible and $h' = h$ is a λ-o.s. tensor norm, the conclusion follows.

(2) The mapping ideals $\mathcal{N}$ and $\mathcal{I}$ are right-accessible and locally left-accessible. The mapping ideal $\mathcal{I}$ is not left-accessible.

From the fact that proj is both right and left accessible we obtain the first assertion. Now, suppose that $\mathcal{I}$ is left-accessible. Then, for every $E \in \mathrm{OBAN}$, $N \in \mathrm{OFIN}$, $T \in \mathrm{CB}(E, N)$ and $\varepsilon > 0$, there exist $L \in \mathrm{OCOFIN}(E)$ and $S \in \mathrm{CB}(E/L, N)$ such that $T = S \circ q_L^E$ and $\iota(S) \leq (1 + \varepsilon)\iota(T)$. Due to the fact that $E/L, N \in \mathrm{OFIN}$ we know that $\nu(T) \leq \nu(S) = \iota(S)$ which implies $\nu(T) \leq \iota(T)$. Since the other inequality is always valid we obtain the isometry $\mathcal{N}(E, N) = \mathcal{I}(E, N)$. By [34, Prop. 14.3.1] this isometry is valid for every $N \in \mathrm{OFIN}$ if and only if E is locally reflexive. Therefore, the mapping ideal $\mathcal{I}$ can not be left-accessible.

(3) The mapping ideal $\mathfrak{A}_{h+h^t}$ is totally accessible and the mapping ideal $\mathfrak{A}_{h\cap h^t}$ is accessible.

   The norm $h \cap h^t$ is totally accessible, the norm $h + h^t$ is accessible and both dual norms are $\mathcal{E}(\lambda)$-o.s. tensor norms.

We are not aware of whether there exists a locally right-accessible mapping ideal which is not right-accessible. It would be interesting to have such an example or a proof of its nonexistence.

Chapter 12
Natural Operator Space Tensor Norms

Grothendieck's *Résumé* contained the list of all *natural* tensor norms. These norms come from applying a finite number of *natural* operations to the projective and injective tensor norms. They are obtained by taking left/right projective and injective hulls in some order (see Sections 15 and 20 in [23]). Grothendieck proved that there were at most fourteen possible natural norms, but he did not know the exact dominations among them, or if there was a possible reduction on the table of natural norms (this was, in fact, one of the open problems posed in the *Résumé*). This was solved, several years later, thanks to very deep ideas of Gordon and Lewis. All these results are now classical and can be found for example in [23, Sec. 27] and [27, 4.4.2].

One of the strengths of Grothendieck's result is that most of his fourteen tensor norms (at least ten) are *really natural*, since they turn out to be equivalent to the most relevant tensor norms: those related to the ideals of bounded, integral, absolutely r-summing ($r = 1, 2$), r-factorable ($r = 1, 2, \infty$) and 2-dominated operators. These tensor norms appear *naturally* in the theory by their own interest, and it is a remarkable thing that they can be obtained from the projective/injective norm by means of the *natural* operations introduced by Grothendieck.

In the operator space setting a study of natural norms in the sense of Grothendieck was suggested by Blecher in [10] while in [48, 68] completely injective and completely projective hulls of tensor norms are considered in the C^*-algebra framework. However, only a few natural operator space tensor norms have been previously considered as related to mapping ideals. That is the case of $\min /$, $\setminus \min$ and $/ \operatorname{proj} \setminus$. According to Example 10.3.13 we have that $d_\infty = \min /$, and this was also observed in [33] (right after Cor. 5.5). By transposing, $g_\infty = \setminus \min$. The o.s. tensor norm $/ \operatorname{proj} \setminus$ was introduced in [29] under the name η and the dual of $E \otimes_\eta F$ can be identified with the operator space bilinear ideal of extendible elements (i.e. those which extend to any larger space). As can be seen in [29] the o.s. tensor norm $/ \operatorname{proj} \setminus$ is connected to deep results on noncommutative versions of Grothendieck's inequality [42, 69], and it actually is equivalent at the Banach

© The Author(s), under exclusive license to Springer Nature Switzerland AG 2025
J. A. Chávez-Domínguez et al., *Operator Space Tensor Norms*, Lecture Notes
in Mathematics 2379, https://doi.org/10.1007/978-3-031-96209-7_12

space level to $h \cap h^t$. Similarly $\backslash \min /$ is equivalent to $h + h^t$ at the Banach space level (see Lemma 12.4.1 for the details).

The lack of a proper tensor product version of Grothendieck's inequality along with (or maybe motivated by) the existence of Haagerup tensor norm which is both completely injective and completely projective has a strong impact in the behavior of *natural* operator space tensor norms.

In this chapter, we consider Grothendieck's natural norms in the operator space framework. In other words, those norms obtained from min or proj after applying left or right injective/projective hulls finitely many times.

We present a first overview of the list, proving many dominations and non-equivalences. We also state an interesting list of open questions about them as well. On the positive side, we completely describe the list of all natural norms that come from applying to min or proj two-sided symmetric operations (injective or projective hulls). Precisely, the list consists of six o.s. tensor norms. Again, this differs from the Banach space case where there are four.

12.1   Similarities and Differences with the Banach Space Framework

We begin with some results from Banach space theory that remain valid for operator spaces and later we get into the differences between both contexts.

As in the Banach space setting the process of taking alternatively injective and projective hulls (at the same side or at both sides of a norm) finishes after three steps. This is shown in the following lemma which is the o.s.version of [27, Prop. 2.6.3] (see also [16, Lem. 3.3]).

Lemma 12.1.1 *Let α be an o.s. tensor norm then*

(a) $\backslash(/(\backslash(/\alpha\backslash)/)\backslash)/ = \backslash(/\alpha\backslash)/$ *and* $/(\backslash(/(\backslash\alpha/)\backslash)/)\backslash = /(\backslash\alpha/)\backslash$.
(b) $\alpha\backslash/\backslash/ = \alpha\backslash/$ *and* $\alpha/\backslash/\backslash = \alpha/\backslash$.
(c) $\backslash/\backslash/\alpha = \backslash/\alpha$ *and* $/\backslash/\backslash\alpha = /\backslash\alpha$.

Proof We just prove the first equality of *(a)* since all the arguments are similar. It is clear that $/(\backslash(/\alpha\backslash)/)\backslash \leq \backslash(/\alpha\backslash)/$ so we can apply completely projective hulls to both sides to obtain $\backslash(/(\backslash(/\alpha\backslash)/)\backslash)/ \leq \backslash(/\alpha\backslash)/$. The other inequality follows analogously: we start from $/\alpha\backslash \leq \backslash(/\alpha\backslash)/$ and apply completely injective hulls to both sides and then completely projective hulls arriving to $\backslash(/\alpha\backslash)/ \leq \backslash(/(\backslash(/\alpha\backslash)/)\backslash)/$. $\qquad\square$

As we have seen in Definitions 10.2.3 and 10.3.10 we can commute the order when taking completely injective (or completely projective) hulls on both sides of a norm. The situation is different if we apply a completely injective hull on one side and a completely projective hull on the other. The following lemma (which has an analogous statement for the Banach space setting) provides an example of this fact through an equality of norms which will be important for the future description of natural norms.

Lemma 12.1.2 *The following o.s. tensor norms are equal:*

$$(/\operatorname{proj})/ = \operatorname{proj} = \backslash(\operatorname{proj}\backslash).$$

Proof By Lemmas 10.2.2 and 10.3.7, $(/\operatorname{proj})/$ is finitely-generated. So, to prove the first equality (the other is obtained by transposing) it is enough to work with $E, F \in$ OFIN. Let $q_F : Z_F \twoheadrightarrow F$ be a complete quotient. Then the following two mappings are complete quotients too:

$$id_E \otimes q_F : E \otimes_{/\operatorname{proj}} Z_F \twoheadrightarrow E \otimes_{(/\operatorname{proj})/} F \quad \text{and} \quad id_E \otimes q_F : E \otimes_{\operatorname{proj}} Z_F \twoheadrightarrow E \otimes_{\operatorname{proj}} F.$$

Since $E \otimes_{/\operatorname{proj}} Z_F = E \otimes_{\operatorname{proj}} Z_F$ (because Z'_F is completely injective -see Remark 12.2.2 below we deduce that $E \otimes_{(/\operatorname{proj})/} F = E \otimes_{\operatorname{proj}} F$. $\qquad\square$

The min counterpart of the previous lemma (which is valid in the Banach space framework) is not true in the operator space world. To prove this we appeal to the following result which can be also obtained (after dualization) as a byproduct of the proof of [34, Prop. 15.4.3] together with the fact that the set of exactly integral mappings is contained in the set of completely 1-summing mappings.

Lemma 12.1.3 *Let $E \in$ OBAN and $F \in$ OFIN with $F' \subset E$. If there exists $C > 0$ such that $\|id : E \otimes_{\min} F \to E \otimes_{\min/} F\| \leq C$ then the operator space F' is C-exact.*

Proof Since F is finite-dimensional, there is a canonical completely isometric isomorphism $\Phi : E \otimes_{\min} F \to \operatorname{CB}(F', E)$. Let $z \in E \otimes_{\min} F$ such that $\Phi(z) = id$, which leads to $\min(z; E, F) = 1$ and $\min/(z; E, F) \leq C$. Given $\varepsilon > 0$, by Corollary 10.3.14 there exist $k \in \mathbb{N}$, $R : S_1^k \to F$ and $u \in E \otimes_{\min} S_1^k$ such that $(id \otimes R)(u) = z$, $\|R\|_{\operatorname{cb}} \leq 1 + \varepsilon$ and

$$\min(u; E, S_1^k) \leq (1 + \varepsilon)\min/(z; E, F) \leq C(1 + \varepsilon).$$

Translating this through Φ we obtain the following commutative diagram:

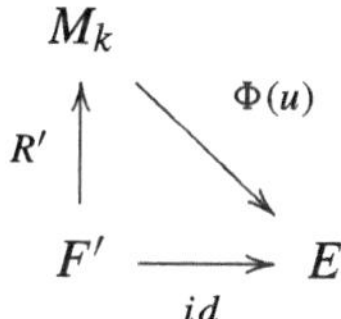

where $\|R'\|_{\operatorname{cb}} \leq 1 + \varepsilon$ and $\|\Phi(u)\|_{\operatorname{cb}} \leq C(1 + \varepsilon)$. Hence, F' is C-exact. $\qquad\square$

The statement of the next lemma is (maybe) a little bit stronger than the one we have promised. We show that the norms $/(\min/\backslash)$ and $(/\backslash\min)\backslash$ differ from $\min$. Since $/(\min/\backslash) \leq /(\min/)$ and $(/\backslash\min)\backslash \leq (\backslash\min)\backslash$ the expected result follows.

Lemma 12.1.4 *Neither* $/(\min/\backslash)$ *nor* $(/\backslash\min)\backslash$ *are equivalent to* $\min$. *In particular, neither* $/(\min/)$ *nor* $(\backslash\min)\backslash$ *are equivalent to* $\min$.

Proof Let us consider the operator space S_1^n for any $n > 2$ and a Hilbert space H such that $S_1^n \subset \mathcal{B}(H)$. Recall that $S_1^n = M_n'$.

Since $\mathcal{B}(H)$ and M_n are completely injective we know that $\mathcal{B}(H) \otimes_{/(\min/\backslash)} M_n = \mathcal{B}(H) \otimes_{\min/} M_n$. Thus, to prove that $/(\min/\backslash)$ and $\min$ are not equivalent it is enough to see that $\mathcal{B}(H) \otimes_{\min/} M_n$ and $\mathcal{B}(H) \otimes_{\min} M_n$ are not uniformly completely isomorphic. By Lemma 12.1.3 this is true since there is no C such that S_1^n is C-exact for all n [34, Thm. 14.5.4]. $\square$

Remark 12.1.5 From Lemmas 12.1.2 and 12.1.4 we derive several consequences:

(a) The o.s. tensor norm $\min/$ is not left-accessible (and the o.s. tensor norm $\backslash\min$ is not right-accessible).

   Indeed, since we know from Example 11.1.3 that $(\min/)' = \text{proj}\backslash$, if $\min/$ was left-accessible by appealing to Proposition 11.1.1 we would have that $/(\min/) = (\backslash(\text{proj}\backslash))' = \text{proj}' = \min$ which is not true.

(b) If $\alpha = \min/$ and $\beta = \backslash\min$ since both are finitely-generated the previous item shows that

$$(/\alpha)' \neq \backslash\alpha' \quad \text{and} \quad (\beta\backslash)' \neq \beta'/.$$

Also, note that $/(\min/\backslash) = /(\min/)\backslash$ and $(/\backslash\min)\backslash = /(\backslash\min)\backslash$. Then, by Lemma 12.1.4 and observing that $\backslash(\text{proj}\backslash)/ = \backslash(/\text{proj})/ = \text{proj}$ we obtain

$$(/\alpha\backslash)' \neq \backslash\alpha'/ \quad \text{and} \quad (/\beta\backslash)' \neq \backslash\beta'/.$$

(c) The o.s. tensor norm $\gamma = /(\min/)$ is completely left-injective, but its dual γ' is not completely left-projective. Analogously, the o.s. tensor norm $(\backslash\min)\backslash$ is completely right-injective but its dual is not completely right-projective. Also, the o.s. tensor norms $/(\backslash\min)\backslash$ and $/(\min/)\backslash$ are (both sides) completely injective but their duals are not (both sides) completely projective .

   Indeed, since $\min \leq \gamma \leq \min/$ we have that $\text{proj} \geq \gamma' \geq \text{proj}\backslash$. If γ' were completely left-projective we would have $\text{proj} \geq \gamma' \geq \backslash(\text{proj}\backslash) = \text{proj}$ which is not true.

The previous remark shows that the dual of a completely injective operator space tensor norm is not necessarily completely projective, exhibiting that the definition of projective hulls (as the dual of the injective hull) given in [10] is not adequate.

As in the Banach space setting we obtain a smaller norm if we apply first a completely projective hull on one side and then the completely injective hull on the other, than if we do so the other way around.

Lemma 12.1.6 *Let α be an o.s. tensor norm. Then:*

$$/(\alpha/) \leq (/\alpha)/ \quad and \quad (\backslash\alpha)\backslash \leq \backslash(\alpha\backslash).$$

Proof We have just to prove the first inequality because the other is obtained by transposing. Due to Remark 10.3.5 along with a slight refinement of Lemma 10.2.2, both norms $/(\alpha/)$ and $(/\alpha)/$ are finitely-generated from the right. Thus, it is enough to check the inequality for $E \otimes F$ with $E \in$ ONORM and $F \in$ OFIN. Consider a complete isometry $i_E : E \hookrightarrow \mathcal{B}(H_E)$ and a complete quotient $q_F : Z_F \twoheadrightarrow F$.

By the projectivity of Z_F, the mapping $id_{\mathcal{B}(H_E)} \otimes q_F : \mathcal{B}(H_E) \otimes_\alpha Z_F \to \mathcal{B}(H_E) \otimes_{\alpha/} F$ is a complete quotient; hence a complete contraction. This implies (by the injectivity of $\mathcal{B}(H_E)$) that $id_E \otimes q_F : E \otimes_{/\alpha} Z_F \to E \otimes_{/(\alpha/)} F$ is a complete contraction.

Now the conclusion follows from the following commutative diagram:

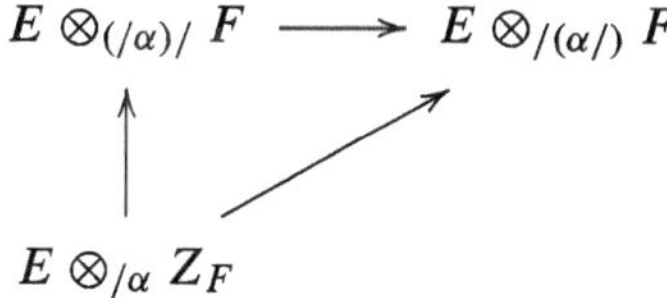

Note that the inequalities of the previous lemma are not in general equalities. Indeed, $/(\text{proj}/) = /\text{proj}$ is not equivalent to proj. So, from Lemma 12.1.2 we derive $/(\text{proj}/) \neq (/\text{proj})/$ and $(\backslash\text{proj})\backslash \neq \backslash(\text{proj}\backslash)$.

The same happens in the Banach space setting: $/(\pi/) \neq (/\pi)/$ and $(\backslash\pi)\backslash \neq \backslash(\pi\backslash)$. Be aware of a false assertion about this topic given in [27, Cor. 2.4.17] (which is deduced from the erroneous statement of [27, Prop. 2.4.16]).

Let us denote by "proj *family*" (resp. "min *family*") the set of all the o.s. tensor norms produced by applying one-sided completely injective and/or completely projective hulls any number of times to the norm proj (resp. min). Note that if we translate this to the Banach space context we obtain that the union of both families is the set of Grothendieck's natural tensor norms.

It is clear that $/\text{proj}\backslash$ is the largest completely injective o.s. tensor norm and $\backslash\text{min}/$ is the smallest completely projective o.s. tensor norm. The following simple lemma says that they are also the smallest and the largest members of their respective families.

Lemma 12.1.7

(a) *The smallest norm of the* proj *family is* $/\text{proj}\backslash$.
(b) *The largest norm of the* min *family is* $\backslash\text{min}/$.

Proof The proof of (*a*) follows easily from the following self explanatory fact: if $/\operatorname{proj}\backslash \leq \alpha$ for some o.s. tensor norm α, then

$$/\operatorname{proj}\backslash \leq \alpha\backslash, \quad /\operatorname{proj}\backslash \leq /\alpha, \quad /\operatorname{proj}\backslash \leq \alpha/, \quad /\operatorname{proj}\backslash \leq \backslash\alpha.$$

The proof of (*b*) is similar.   □

Remark 12.1.8 (Some Tensor Norm Consequences of Grothendieck's Inequality that Are Not Longer Valid in the Operator Space Framework)

(a) It follows from Grothendieck's inequality that $/\pi\backslash$ is dominated by $\backslash\varepsilon/$ (i.e. there exists a constant $C > 0$ such that $/\pi\backslash \leq C\backslash\varepsilon/$).

   In the operator space setting, it is not possible for a member of the min family to dominate a member of the proj family. Indeed, since the Haagerup tensor norm h is both completely injective and completely projective we have that

$$\backslash\min/ \leq h \leq /\operatorname{proj}\backslash.$$

   The lack of symmetry of h prevents it from being equivalent to any symmetric tensor norm (as $\backslash\min/$ or $/\operatorname{proj}\backslash$). To check this recall [34, Prop. 9.3.4] that for any Hilbert space H if we denote by H_c the associated column space we have that $H_c \otimes_h H'_c = \mathcal{K}(H)$ while $H'_c \otimes_h H_c = \mathcal{B}(H)$.

   By means of the previous lemma, we get that any member of the min family is smaller than (and not equivalent to) any member of the proj family.
(b) As a byproduct of the inequality referred in (a) it is obtained that the norms $\backslash(/\pi\backslash)/$ and $\backslash\varepsilon/$ are equivalent; the same happens with $/(\backslash\varepsilon/)\backslash$ and $/\pi\backslash$.

   The operator space version of these equivalences is clearly not valid due to the arguments explained in the previous item. This fact certainly has an impact in the quantity of *natural* operator space tensor norms.
(c) In the Banach space world the equivalences stated in the previous item obviously produce that if α is an injective tensor norm then $\backslash\alpha/$ is equivalent to $\backslash\varepsilon/$ and if α is a projective tensor norm then $/\alpha\backslash$ is equivalent to $/\pi\backslash$.

   In the operator space universe these equivalences are obviously not true. From our explanation in (a) it is clear that $\backslash\min/ \leq h \leq \backslash(/\operatorname{proj}\backslash)/$ are three non-equivalent completely projective hulls of completely injective norms, and $/(\backslash\min/)\backslash \leq h \leq /\operatorname{proj}\backslash$ are three non-equivalent completely injective hulls of completely projective norms.

Remark 12.1.9 (Open Question) In the Banach space setting the impact of Grothendieck's inequality in the description of natural tensor norms is decisive. We have pointed out in the previous remark several relations that are not longer valid in the operator space framework.

   There is another equivalence of Banach space tensor norms (also derived from Grothendieck's inequality) that is crucial for the resulting number of 14 natural tensor norms, namely:

The norms $/((/\pi\backslash)/)$ and $(/\pi\backslash)/$ are equivalent.

To recall the sketch of the argument we use the symbol $\sim$ to denote equivalent tensor norms and we denote by w_2 the Hilbertian tensor norm and by d_2^B the Chevet-Saphar tensor norm. By Grothendieck's inequality, $/\pi\backslash \sim w_2$ and $w_2/ \sim d_2^B$. Since d_2^B is completely left-injective the desired equivalence is proved.

In the operator space setting we cannot follow the same path, but nevertheless we can ask:

Are the norms $/((/\operatorname{proj}\backslash)/)$ and $(/\operatorname{proj}\backslash)/$ equivalent?

Observe that clearly $/((/\operatorname{proj}\backslash)/) \leq (/\operatorname{proj}\backslash)/$ so the question can be equivalently posed in the following way:

Is the norm $(/\operatorname{proj}\backslash)/$ completely left-injective?

If the answer to the open question is positive we obtain a large number of coincidences between members of the proj family. Indeed, an affirmative answer would turn into identities the following inequalities derived from Lemma 12.1.6:

(a) $/(\operatorname{proj}\backslash/) \leq (/\operatorname{proj}\backslash)/$ and $(\backslash/\operatorname{proj})\backslash \leq \backslash(/\operatorname{proj}\backslash)$.
(b) $\backslash/(\operatorname{proj}\backslash/) \leq \backslash(/\operatorname{proj}\backslash)/$ and $(\backslash/\operatorname{proj})\backslash/ \leq \backslash(/\operatorname{proj}\backslash)/$.

Also, if there are identities in (a) the following inequalities turn out also to be identities:

(c) $/\operatorname{proj}\backslash \leq (/\operatorname{proj}\backslash)/\backslash$ and $/\operatorname{proj}\backslash \leq /\backslash(/\operatorname{proj}\backslash)$.

12.2 The proj Family

Even though we are not able to give a full description of the proj family, we devote this Subsection to present a picture of what we know.

The proj family begins of course with the proj norm. Then, we enumerate the norms taking into account how many procedures (of completely left or completely right-injective/projective hull) we have applied to proj in order to obtain them. The list, *up to four procedures* is the following:

(0) proj.
(1) $/\operatorname{proj} \bullet \operatorname{proj}\backslash$.
(2) $/\operatorname{proj}\backslash \bullet \backslash/\operatorname{proj} \bullet \operatorname{proj}\backslash/$.
(3) $\backslash(/\operatorname{proj}\backslash) \bullet (/\operatorname{proj}\backslash)/ \bullet (\backslash/\operatorname{proj})\backslash \bullet /(\operatorname{proj}\backslash/)$.
(4) $\backslash(/\operatorname{proj}\backslash)/ \bullet /\backslash(/\operatorname{proj}\backslash) \bullet (/\operatorname{proj}\backslash)/\backslash \bullet (\backslash/\operatorname{proj})\backslash/ \bullet \backslash/(\operatorname{proj}\backslash/)$.

The norms (derived from proj after at most four procedures) that do not appear in the previous list are identical to one of those that do appear in the list. Indeed, from Lemmas 12.1.2, 12.1.6, 12.1.7 and the definitions of completely injective and completely projective hulls we have:

(2) $(/\operatorname{proj})/ = \backslash(\operatorname{proj}\backslash) = \operatorname{proj}$.

(3) $(\backslash/\mathrm{proj})/ = \backslash(\mathrm{proj}\,\backslash/)) = \mathrm{proj}.$
(4) $/((/\mathrm{proj}\,\backslash)/) = /(\mathrm{proj}\,\backslash/) \bullet (\backslash(/\mathrm{proj}\,\backslash))\backslash = (\backslash/\mathrm{proj})\backslash.$
(4) $(/(\mathrm{proj}\,\backslash/))/ = (/\mathrm{proj}\,\backslash)/ \bullet \backslash((\backslash/\mathrm{proj})\backslash) = \backslash(/\mathrm{proj}\,\backslash).$
(4) $/(\backslash/\mathrm{proj})\backslash = /(\mathrm{proj}\,\backslash/)\backslash = /\,\mathrm{proj}\,\backslash.$

Thus, the members of the proj family *up to four procedures* are at most 15. We do not know whether they are all non-equivalent. In fact, as we mentioned above, if the answer of the open question is positive we derive six identities between members of the proj family, reducing the list from 15 to 9 members.

Moreover, an affirmative answer to the open question would also imply that there exist no more members of the proj family: a fifth procedure produces no new elements. Indeed, in this situation, the fourth procedure only generates one element: $\backslash(/\mathrm{proj}\,\backslash)/ = (\backslash/\mathrm{proj})\backslash/ = \backslash/(\mathrm{proj}\,\backslash/)$. And it is easily seen that a completely left or completely right-injective hull of this element (under the present hypothesis of an affirmative answer to the open question) yields the norm $/\,\mathrm{proj}\,\backslash$.

Now, we present the dominations valid between the 15 norms of our list and later we will provide examples to show that many of these norms are non-equivalent. Each arrow $\alpha \to \beta$ in the next diagram means that $\beta \le \alpha$. Double arrows are those where we prove in the sequel that the dominations are strict. Tensor norms connected by doted arrows are equivalent if the Open question Remark 12.1.9 has an affirmative answer. We do not know if the dominations connected by standard arrows are in fact strict.

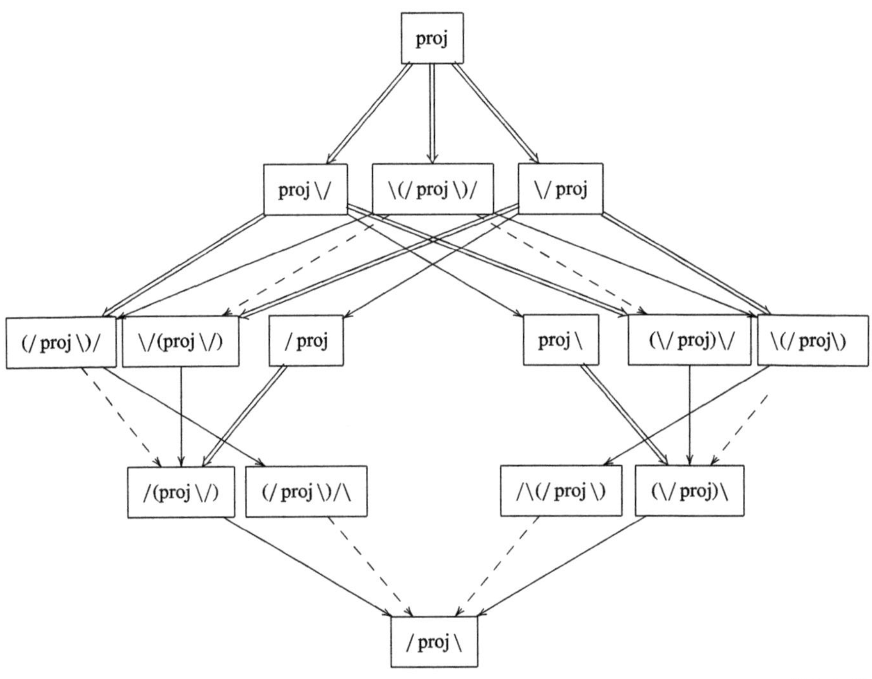

$$(12.2.1)$$

The dominations stated in the previous diagram are easily derived from the definitions of completely injective and completely projective hulls plus the results already proved in Lemmas 12.1.2, 12.1.6 and 12.1.7.

The following two remarks look into representations of elements belonging to the dual of tensor products with particular tensor norms. They will be useful to show that some of the previous dominations can not be turned into equivalences.

Remark 12.2.1 If $T \in \mathrm{CB}(E, F')$ belongs to $(E \otimes_{/\mathrm{proj}\backslash} F)'$ and $i_E : E \to \mathcal{B}(H_E)$ and $i_F : F \to \mathcal{B}(H_F)$ are complete isometries then there exists a completely bounded mapping $\widetilde{T} : \mathcal{B}(H_E) \to \mathcal{B}(H_F)'$ such that the following diagram commutes:

$$
\begin{array}{ccc}
E & \xrightarrow{\;\;T\;\;} & F' \\[2pt]
{\scriptstyle i_E}\downarrow & & \uparrow{\scriptstyle i'_F} \\[2pt]
\mathcal{B}(H_E) & \xrightarrow[\;\widetilde{T}\;]{} & \mathcal{B}(H_F)'
\end{array}
$$

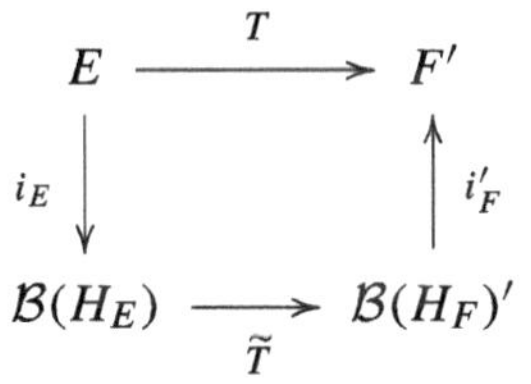

Appealing to [66, Thm. 18.1] we know that $\widetilde{T}$ (and hence T) is weakly compact.

Remark 12.2.2 If F' is a completely injective operator space then $E \otimes_{/\mathrm{proj}} F = E \otimes_{\mathrm{proj}} F$.

Indeed, for $T \in \mathrm{CB}(E, F')$ we have that $T \in (E \otimes_{/\mathrm{proj}} F)'$ if for any complete isometry $i_E : E \to \mathcal{B}(H_E)$ there exists a completely bounded mapping $\widetilde{T} : \mathcal{B}(H_E) \to F'$ such that the following diagram commutes:

$$
\begin{array}{ccc}
E & \xrightarrow{\;\;T\;\;} & F' \\[2pt]
{\scriptstyle i_E}\downarrow & \nearrow{\scriptstyle \widetilde{T}} & \\[2pt]
\mathcal{B}(H_E) & &
\end{array}
$$

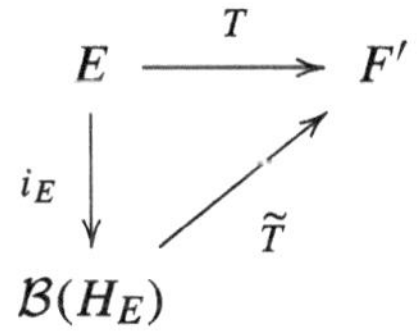

The injectivity of F' implies that this happens to any mapping $T \in \mathrm{CB}(E, F')$ (and that the norm is maintained).

Of course, an analogous result holds for the completely right-injective hull: if E' is a completely injective operator space then $E \otimes_{\mathrm{proj}\backslash} F = E \otimes_{\mathrm{proj}} F$.

Example 12.2.3 The following pairs of norms are not equivalent: proj and $\backslash/$ proj; proj and proj $\backslash/$; proj $\backslash$ and $(\backslash/ \mathrm{proj})\backslash$; $/$ proj and $/(\mathrm{proj} \backslash/)$.

We carry out the argument for the first and third pairs of norms. The other two are obtained by transposing.

Let us consider the completely projective space $Z = \ell_1(\{S_1^n : n \in \mathbb{N}\})$ and recall that its dual $Z' = \ell_\infty(\{M_n : n \in \mathbb{N}\})$ is an injective space. Then we have the following identities:

$$Z \otimes_{\text{proj}} Z' = Z \otimes_{\text{proj}\backslash} Z' \quad \text{and}$$

$$Z \otimes_{(\backslash/\,\text{proj})\backslash} Z' = Z \otimes_{\backslash/\,\text{proj}} Z' = Z \otimes_{/\,\text{proj}} Z' = Z \otimes_{/\,\text{proj}\backslash} Z'.$$

Thus, the desired results are proved if we show that $Z \otimes_{\text{proj}} Z' \neq Z \otimes_{/\,\text{proj}\backslash} Z'$.

Let $\kappa_Z \in \text{CB}(Z, Z'')$ be the canonical inclusion. Then, $\kappa_Z \in (Z \otimes_{\text{proj}} Z')'$. Now, Remark 12.2.1 says that if $\kappa_Z \in (Z \otimes_{/\,\text{proj}\backslash} Z')'$ then κ_Z would be weakly compact. Since this is not possible because Z is not reflexive, we deduce that

$$\kappa_Z \in (Z \otimes_{\text{proj}} Z')' \setminus (Z \otimes_{/\,\text{proj}\backslash} Z')'$$

and hence $Z \otimes_{\text{proj}} Z' \neq Z \otimes_{/\,\text{proj}\backslash} Z'$.

Example 12.2.4 The following pairs of norms are not equivalent: $\backslash/\,\text{proj}$ and $\backslash(/\,\text{proj}\,\backslash)$; $\text{proj}\,\backslash/$ and $(/\,\text{proj}\,\backslash)/$.

Let us consider again the completely projective space $Z = \ell_1(\{S_1^n : n \in \mathbb{N}\})$ and let $i_Z : Z \hookrightarrow \mathcal{B}(H)$ be a complete isometry, for a suitable Hilbert space H. Let us recall that $S_1(H)$ is an operator space pre-dual of $\mathcal{B}(H)$, and the latter is a completely injective space. Now we have, by Remark 12.2.2,

$$Z \otimes_{\text{proj}} S_1(H) = Z \otimes_{/\,\text{proj}} S_1(H) = Z \otimes_{\backslash/\,\text{proj}} S_1(H) \quad \text{and}$$

$$Z \otimes_{/\,\text{proj}\backslash} S_1(H) = Z \otimes_{\backslash(/\,\text{proj}\,\backslash)} S_1(H).$$

It is clear that $i_Z \in (Z \otimes_{\text{proj}} S_1(H))'$ but it cannot belong to $(Z \otimes_{/\,\text{proj}\backslash} S_1(H))'$ because it is not weakly compact (since it is an isometry from a non reflexive space).

Example 12.2.5 The following pairs of norms are not equivalent: proj and $\backslash(/\,\text{proj}\,\backslash)/$; $\backslash/\,\text{proj}$ and $\backslash/(\text{proj}\,\backslash/)$; $\text{proj}\,\backslash/$ and $(\backslash/\,\text{proj})\backslash/$.

We consider the Banach space $\ell_1(\mathbb{Z})$ with the maximal operator space structure. This is a completely projective space (being an ℓ_1-sum of the projective space $\mathbb{C}$) and thus its dual is completely injective [65, Cor. 24.6]. Then, due to Remark 12.2.2,

$$\ell_1(\mathbb{Z}) \otimes_{\text{proj}} \ell_1(\mathbb{Z}) = \ell_1(\mathbb{Z}) \otimes_{\text{proj}\backslash} \ell_1(\mathbb{Z}) = \ell_1(\mathbb{Z}) \otimes_{\text{proj}\backslash/} \ell_1(\mathbb{Z}).$$

Also we have

$$\ell_1(\mathbb{Z}) \otimes_{\backslash(/\,\text{proj}\,\backslash)/} \ell_1(\mathbb{Z}) = \ell_1(\mathbb{Z}) \otimes_{/\,\text{proj}\backslash} \ell_1(\mathbb{Z}).$$

Now we can borrow an example from Banach space theory [78, Ex. 1.1]. Let $T : \ell_1(\mathbb{Z}) \to \ell_\infty(\mathbb{Z})$ be given by

$$T(x) = \left(\sum_{m \in \mathbb{Z}} \mathrm{sgn}(m)x_{m-n}\right)_{n \in \mathbb{N}}.$$

Then $T \in (\ell_1(\mathbb{Z}) \otimes_{\mathrm{proj}} \ell_1(\mathbb{Z}))'$ (because in $\ell_1(\mathbb{Z})$ with the maximal operator space structure the proj tensor product coincides with the Banach π tensor product).

Once again we have that $T \notin (\ell_1(\mathbb{Z}) \otimes_{/\mathrm{proj}\backslash} \ell_1(\mathbb{Z}))'$ because T is not weakly compact [78, page 50].

12.3   The min Family

Surprisingly the min family is not the *mirror reflection* of the proj family as it happens in the Banach space setting. The reason, of course, comes from Lemmas 12.1.2 and 12.1.4.

Due to this fact there are probably more members in the min family than in the proj family. Hence, in order to have a manageable set, we chose to present a diagram of the min family just *up to three procedures*.

The list is the following:

(0)   min.
(1) $\backslash$ min $\bullet$ min $/$.
(2) $\backslash$ min $/$ $\bullet$ $/\backslash$ min $\bullet$ min $/\backslash$ $\bullet$ $(\backslash$ min$)\backslash$ $\bullet$ $/($min $/)$.
(3) $/(\backslash$ min $/)$ $\bullet$ $(\backslash$ min $/)\backslash$ $\bullet$ $(/\backslash$ min$)/$ $\bullet$ $\backslash($min $/\backslash)$ $\bullet$ $\backslash/($min $/)$ $\bullet$ $(\backslash$ min$)\backslash/$ $\bullet$ $(/\backslash$ min$)\backslash$ $\bullet$ $/($min $/\backslash)$.

The norms (derived from min after at most three procedures) that do not appear in the previous list are identical to one of those that do appear in the list. Indeed, from Lemma 12.1.2 and the definitions of completely injective and completely projective hulls we have:

(3) $(/($min $/))/ = $ min $/$ $\bullet$ $\backslash(($\backslash$ min$)\backslash) = \backslash$ min.
(3) $(/($min $/))\backslash = /($min $/\backslash)$ $\bullet$ $/(($\backslash$ min$)\backslash) = (/\backslash$ min$)\backslash$.

Thus, the members of the min family *up to three procedures* are at most 16. We do not know whether they are all non-equivalent. As in the diagram for the proj family, each arrow $\alpha \to \beta$ means that $\beta \leq \alpha$ and doubled arrows are the only ones where we can prove that the dominations are strict.

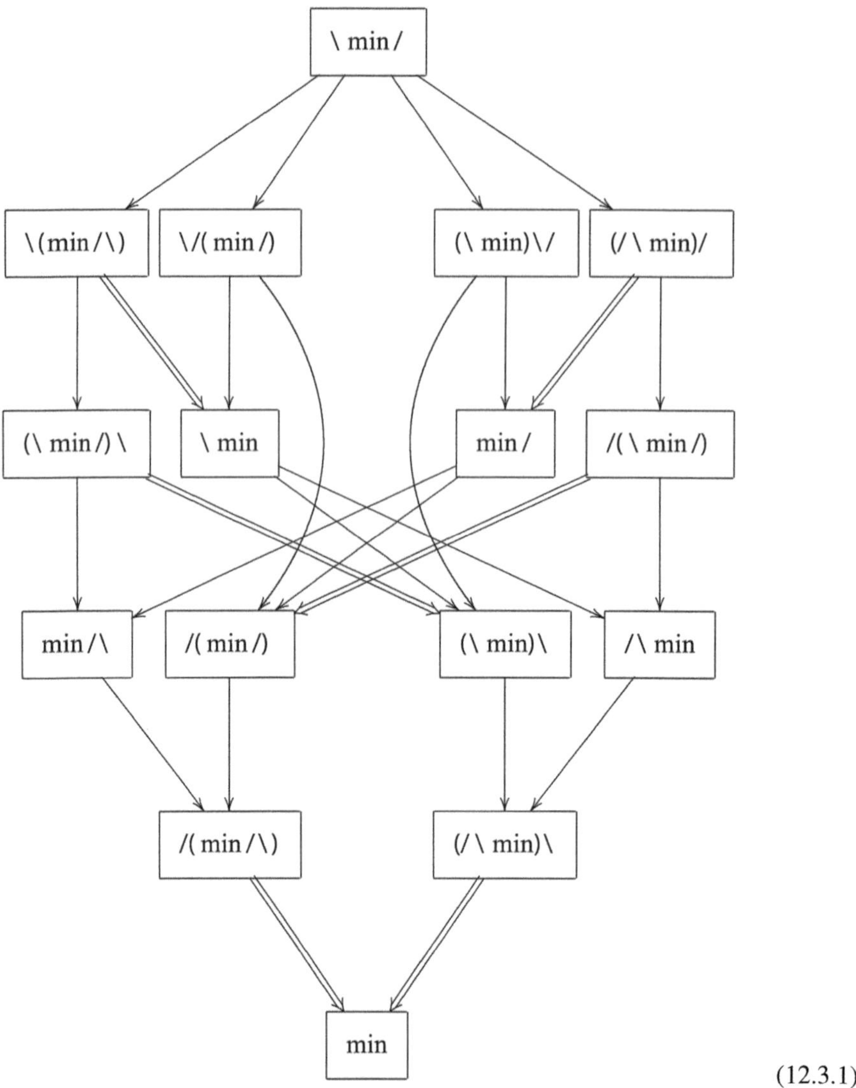

(12.3.1)

Example 12.3.1 The following pairs of norms are not equivalent: $\backslash \min$ and $\backslash(\min /\backslash)$; $(\backslash \min)\backslash$ and $(\backslash \min /)\backslash$; $\min /$ and $(/\backslash \min)/$; $/(\min /)$ and $/(\backslash \min /)$.

As usual, we just argue about the first two non-equivalences. The other two follow by transposition. By Lemma 12.1.3 we know that if $\|id : S_1^n \otimes_{\min} M_n \to S_1^n \otimes_{\min /} M_n\| \leq C$ then S_1^n is C-exact. Since there is no C such that S_1^n is C-exact for every n we derive that $S_1^n \otimes_{\min} M_n$ and $S_1^n \otimes_{\min /} M_n$ are not uniformly completely isomorphic. Since S_1^n is completely projective and M_n is completely injective we have $S_1^n \otimes_{\min} M_n = S_1^n \otimes_{\backslash \min} M_n = S_1^n \otimes_{(\backslash \min)\backslash} M_n$ and $S_1^n \otimes_{\min /} M_n = S_1^n \otimes_{\backslash(\min /\backslash)} M_n = S_1^n \otimes_{(\backslash \min /)\backslash} M_n$ from where the result follows.

12.4   Two-sided Natural Operator Space Tensor Norms

In this section we show that those natural o.s. tensor norms that come from applying to min or proj two-sided hulls operations (injective $/ \cdot \backslash$, or projective $\backslash \cdot /$ hulls) are exactly six. This is somewhat of a surprise: as stated in [10], "there is no reason to suppose that this [process] will not give a large number of inequivalent norms" (due to the lack of a direct analog of Grothendieck's inequality in the o.s. setting). Now, Lemma 12.1.1 (a) shows that there are only 6 possible members of the two-sided family and the dominations between them are easily seen:

$$\min \ \leq \ /(\backslash \min /)\backslash \ \leq \ \backslash \min / \ \leq \ / \operatorname{proj} \backslash \ \leq \ \backslash (/ \operatorname{proj} \backslash)/ \ \leq \operatorname{proj}$$

Also, in Remark 12.1.8 we observe that $\backslash \min /$ and $/ \operatorname{proj} \backslash$ are not equivalent. From Example 12.2.5 (see also diagram (12.2.1)) we know that $\backslash (/ \operatorname{proj} \backslash)/$ and proj are not equivalent. We prove in Lemma 12.1.4 that min and $(/\backslash \min)\backslash$ are not equivalent. Since $(/\backslash \min)\backslash \ = \ /(\backslash \min)\backslash \ \leq \ /(\backslash \min /)\backslash$ we obtain that min and $/(\backslash \min /)\backslash$ could not be equivalent either.

To conclude that there are six two-sided natural norms all we need to show is the non-equivalence of the following two pairs of norms: $/(\backslash \min /)\backslash$ and $\backslash \min /$; $\backslash (/ \operatorname{proj} \backslash)/$ and $/ \operatorname{proj} \backslash$. In the Banach space setting this kind of statement is clear since a projective norm could not be equivalent to an injective one. In the operator space framework this argument is not available, so we need a different proof.

Lemma 12.4.1 *The norms $/ \operatorname{proj} \backslash$ and $\backslash \min /$ are equivalent at the Banach space level to $h \cap h^t$ and $h + h^t$, respectively.*

Proof In [29, Sec. 6] it is showed, as a consequence of results from [69] and [42], that for any operator spaces E and F the spaces $E \otimes_{/ \operatorname{proj} \backslash} F$ and $E \otimes_{h \cap h^t} F$ are isomorphic. By chasing down the constants in the proof one can see that they are universal, so we conclude that $/ \operatorname{proj} \backslash$ is equivalent to $h \cap h^t$ at the Banach space level. Also, [69, Cor. 0.8 (iii)] says that there is an isomorphism between $A' \otimes_{\min} B'$ and $A' \otimes_{h+h^t} B'$ for any C^*-algebras A and B, with constants independent of the spaces involved. Now, if E and F are operator spaces, we have that the completely projective spaces Z_E and Z_F are duals of C^*-algebras (recall that they are defined as ℓ_1-sums of S_1^n spaces, so they are the duals of the corresponding c_0-sums of M_n spaces [65, Sec. 2.6]). Therefore, we have that $E \otimes_{\backslash \min /} F$ and $E \otimes_{h+h^t} F$ are isomorphic with constants independent of E and F. This means that the o.s. tensor norms $\backslash \min /$ and $h + h^t$ are equivalent at the Banach space level.   □

Remark 12.4.2 In Lemma 12.4.1, one cannot get an equivalence between $\backslash \min /$ and $h + h^t$ at the operator space level. Indeed, it is proved in [3, Thm. 1.1] that $S_1 \otimes_{\min} S_1$ and $S_1 \otimes_{h+h^t} S_1$ are not completely isomorphic, which together with Proposition 10.3.12 says that $S_1 \otimes_{\backslash \min /} S_1$ and $S_1 \otimes_{h+h^t} S_1$ are not completely isomorphic. The aforementioned result [3, Thm. 1.1] can be understood as saying that the isomorphism $\ell_1 \otimes_\varepsilon \ell_1 \cong \ell_1 \otimes_{\gamma_2^*} \ell_1$, which is a formulation of Grothendieck's inequality, does not naturally extend to the operator space setting. Of course there is no mystery in replacing ℓ_1 and $\otimes_\varepsilon$ by S_1 and $\otimes_{\min}$, respectively, but it may not be immediately clear why $h + h^t$ is a sensible replacement for γ_2^* in the operator space context. First, recall that γ_2^* is the norm which is in trace duality with the norm of factorization through a Hilbert space. Since in the operator space setting there are many Hilbertian operator spaces, there would appear to be many possible replacements for γ_2^*. We will not go into the details here, and refer the reader to [3] for more on the history of how the works of Pisier [60] and Haagerup [41] suggest that $h + h^t$ is the "correct" analogue of γ_2^* in the setting of operator algebras. We just point out that as can be seen in Example 8.2.14, the tensor norm $h + h^t$ is indeed in trace duality with a certain type of factorization through Hilbertian operator spaces.

Remark 12.4.3 A close look at the definition of dual tensor norms reveals that if two o.s. tensor norms are equivalent at the Banach space level, then so are their dual o.s. tensor norms (because the norm on the dual of an operator space depends only on the norm of the operator space, and not on the norms of the higher matricial levels). Therefore, recalling that $(h \cap h^t)' = h + h^t$ (see Proposition 5.1.3), as a consequence of the previous lemma we obtain that $(\backslash \min /)'$ is equivalent, at the Banach space level, to $/ \operatorname{proj} \backslash$.

Lemma 12.4.4 *The norms* $/(\backslash \min /)\backslash$ *and* $\backslash \min /$ *are not equivalent. The norms* $\backslash(/ \operatorname{proj} \backslash)/$ *and* $/ \operatorname{proj} \backslash$ *are not equivalent.*

Proof In [53, Rmk. 13] it is proved that $h + h^t$ is not completely injective. In fact, their proof shows more: there is a sequence of pairs of complete isometries $T_n : E_n \to \widetilde{E}_n$ and $R_n : F_n \to \widetilde{F}_n$ so that the inverses of

$$T_n \otimes R_n : E_n \otimes_{h+h^t} F_n \to (T_n \otimes R_n)(E_n \otimes F_n) \subset \widetilde{E}_n \otimes_{h+h^t} \widetilde{F}_n$$

have norms tending to infinity, so $h + h^t$ cannot be equivalent at the Banach space level to a completely injective o.s. tensor norm. As a consequence of Lemma 12.4.1, neither can $\backslash \min /$, and therefore $\backslash \min /$ and $/(\backslash \min /)\backslash$ cannot be equivalent.

Also, we know from Corollary 10.1.4 that the dual of a completely projective tensor norm is completely injective. Thus, if $h \cap h^t$ were equivalent at the Banach space level to a completely projective tensor norm, from the previous remark we would obtain that $(h \cap h^t)' = h + h^t$ would be equivalent at the Banach space level to a completely injective tensor norm. From the previous paragraph we know that this is false, so therefore $/ \operatorname{proj} \backslash$ cannot be equivalent at the Banach space level to a completely projective tensor norm. This, finally, means that $/ \operatorname{proj} \backslash$ and $\backslash(/ \operatorname{proj} \backslash)/$ are not equivalent.   □

Hence, we can represent the two-sided family in the following diagram:

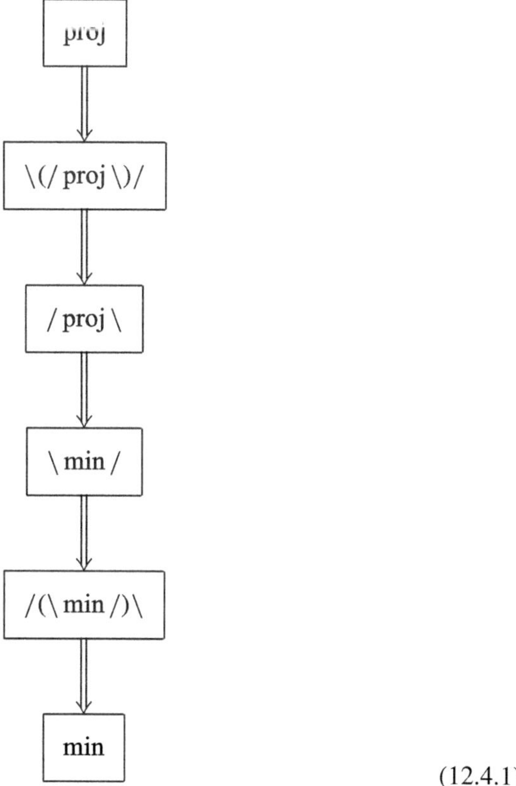

$$(12.4.1)$$

This is different from the Banach space case: those natural norms that come from applying to ε or π two-sided hulls operations are exactly four. This is how the two-sided natural norms in the classical setting are arranged:

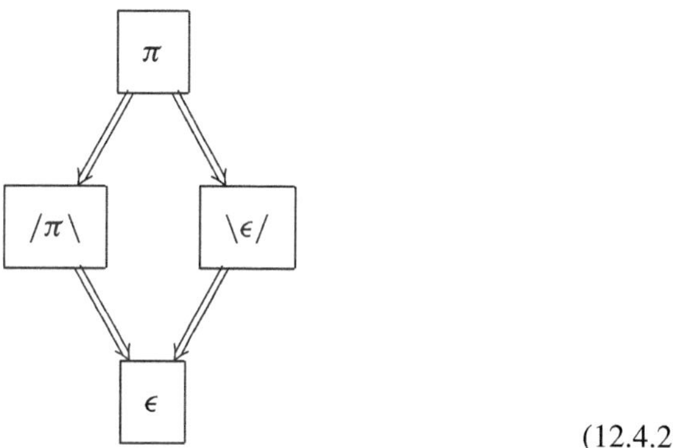

$$(12.4.2)$$

In both diagrams (12.4.1) and (12.4.2) each arrow $\alpha \Rightarrow \beta$ means that $\beta \leq \alpha$ and that these norms are not equivalent. The norms $/\pi\backslash$ and $\backslash\epsilon/$ are not equivalent and there is no domination between them.

Obviously this difference between these frameworks is expected; in the context of operator spaces it is not valid a Grothendieck-type inequality relating the norms as in the classical Banach context: $/\pi\backslash$ and $/(\backslash\epsilon/)\backslash$ are equivalent and also their duals $\backslash\epsilon/$ and $\backslash(/\pi\backslash)/$ while $/\operatorname{proj}\backslash$ and $/(\backslash\min/)\backslash$ are not equivalent, neither $\backslash\min/$ and $\backslash(/\operatorname{proj}\backslash)/$.

Remark 12.4.5 It would be tempting to say that the diagram (12.4.1) shows the complete picture of all **symmetric** natural o.s. tensor norms. While it is clear that we can only get symmetric o.s. tensor norms by starting with min and proj and applying two-sided hulls finitely many times, we do not know whether these are **all** the symmetric natural o.s. tensor norms. For example recall that $(/\operatorname{proj})/ = \backslash(\operatorname{proj}\backslash) = \operatorname{proj}$, so taking hulls in a non-symmetric way can still yield a symmetric o.s. tensor norm.

Chapter 13
Conclusions and Some Open Questions

The main goal of this work is to initiate a program for the theory of tensor products and tensor norms in the category of operator spaces. In particular, we focus on the interplay of these theories and the theory of mapping ideals. Of course, many definitions and results in this framework are natural since they have a corresponding one in the context of Banach spaces. But many of them are not! As noticed many times, the theory is not just a straightforward translation of what is known in the classical setting and new and challenging questions naturally arise. To start with a typical difference, we highlight the role of local reflexivity, which seems to be crucial in many places as is well-known to experts. However, sometimes this hypothesis can be avoided: for example, when dealing with the class of extended λ-o.s. tensor norms. An unexpected issue in this category appears when relating the left accessibility of mapping ideals and associated tensor norms. The fact that their relationship is weaker than the one for its right counterpart is certainly puzzling.

The most surprising finding that we realized during this work is that the min family can not be obtained by dualizing the proj family. This, of course, is very counter intuitive and exhibits, once again, the differences between the classical and the non-commutative theories. All of this is related, in some sense, to the fact that accessibility plays an important role in the duality between injective and projective hulls of a given o.s. tensor norm. This hypothesis does not appear in the theory of Banach spaces. Moreover, these families are completely separate: one family dominates the other. This is because the Haagerup o.s. tensor norm is simultaneously completely injective and projective (an impossible property in the classical realm).

Summarizing, we have begun looking with operator space eyes at some relevant properties and results from the classical theory, and found that to fully develop the theory it is important a different and original perspective, involving novel insights, ideas, techniques or hypotheses. Clearly, there are plenty of unexplored paths for future research. This work aims to contribute to a program that we believe is much vaster. To conclude we want to explicitly recall here some unsolved issues that appeared through this monograph.

J. A. Chávez-Domínguez et al., *Operator Space Tensor Norms*, Lecture Notes in Mathematics 2379, https://doi.org/10.1007/978-3-031-96209-7_13

Problem 1 All along the monograph there are results where some hypotheses about local reflexivity are not needed in the case of $\mathcal{E}(\lambda)$-o.s. tensor norms. It would be welcome to have more examples of λ-o.s. tensor norms, in addition to the typical ones (proj, h and h^t) and the $\odot^\theta$ constructed in Sect. 2.2 in order to expand the profitable family of $\mathcal{E}(\lambda)$-o.s. tensor norms.

Problem 2 We see in Theorem 2.2.7 that for $1 \le p \le q \le \infty$ and for any operator spaces E and F, we have

$$\left\| E \otimes_{d_p} F \to E \otimes_{d_q} F \right\| \le 1.$$

However, we do not know whether this inequality also holds with the completely bounded norm.

Problem 3 We introduce in Chap. 5 the notions of locally right-accessible tensor norm and locally left-accessible tensor norm, but we do not have any example that justifies that these concepts are weaker than their non-local siblings.

Problem 4 It is an interesting question whether there exists a non-accessible o.s. tensor norm. Recall that in the Banach space setting, the existence of such a norm was established using non-trivial arguments (see [23, Th. 31.6]).

Problem 5 We prove in Remark 5.3.10 that the intersection of two cofinitely-generated o.s. tensor norms is also cofinitely-generated. It would be nice to know whether the same is true for the sum procedure. This question for the particular case of $h + h^t$ was posed in Remark 5.3.14.

Problem 6 We consider in Chap. 6 a weak version of the complete bounded approximation property, namely the W*CBAP. We do not know any example of a space without CBAP satisfying W*CBAP. In view of Proposition 6.2.6 such a space can not be locally reflexive.

Problem 7 The theory of α-W*CBAP suggested in Remark 6.2.5 seems to be an interesting topic for future development.

Problem 8 In Chap. 11 it would be interesting to have an example of a locally right-accessible mapping ideal which is not right-accessible. Recall that we show in Example 11.3.6 that $\mathcal{I}$ is an example for the left version of this question.

Problem 9 In Proposition 11.3.4 we see that there is an equivalence between right-accessibility of a mapping ideal and of a tensor norm associated to it. For left-accessibility the result is much weaker and just in one direction. It would be important to know if a statement for the other direction is valid: if the mapping ideal is left-accessible, is it true that the associated tensor norm is (locally) left-accessible?

Problem 10 With respect to natural tensor norms we have barely begun a classification of them and there is a hard road ahead. It is left to be solved whether most of the dominations in diagrams (12.2.1) and (12.3.1) are strict or not.

Problem 11 A solution to Open question Remark 12.1.9 will shed some light on the structure of the proj family after more than 4 procedures.

Problem 12 Our knowledge of the min family is so scarce that almost the whole picture remains to be done.

Problem 13 The fact that the proj and min families are not a reflected copy of one another, along with the necessity of extra conditions to ensure that the dual of the projective hull of a norm is the injective hull of the dual norm, have as a consequence that we do not know which are the duals for most of the natural norms. The main inquiry of this group is whether the following identity holds: $(\setminus \min /)' = /\,\mathrm{proj}\setminus$ (see the comment after Example 11.1.3). Recall, also, that from Remark 12.4.3, $(\setminus \min /)'$ and $/\,\mathrm{proj}\setminus$ are equivalent, at the Banach space level. So, it is expected that they are equivalent, or even equal, as o.s. tensor norms.

Problem 14 Since the min family is certainly larger than the proj family, there should exist norms in the min family whose duals do not belong to the proj family. It would be interesting to identify these norms.

Problem 15 As stated in Remark 12.4.5, it would be nice to know if all the symmetric natural tensor norms are the two-sided natural tensor norms (i.e. those that appear in diagram (12.4.1)).

References

1. Antony, J., Kumar, A.: Spectra of elements in operator space tensor products of C*-algebras. Positivity **25**(5), 1973–1987 (2021). https://doi.org/10.1007/s11117-021-00856-z
2. Antony, J., Kumar, A., Luthra, P.: Operator space tensor products and inductive limits. J. Math. Anal. Appl. **470**(1), 235–250 (2019). https://doi.org/10.1016/j.jmaa.2018.09.067
3. Araiza, R., Junge, M., Palazuelos, C.: On a question of Blecher, Pisier, Shlyakhtenko (2024). https://arxiv.org/abs/2406.05302
4. Arhancet, C.: Entanglement-assisted classical capacities of some channels acting as radial multipliers on fermion algebras. J. Funct. Anal. **288**(5), Paper No. 110790, 47 (2025). https://doi.org/10.1016/j.jfa.2024.110790
5. Arhancet, C.: Quantum information theory and Fourier multipliers on quantum groups (2025). https://arxiv.org/abs/2008.12019
6. Beckner, W.: Inequalities in Fourier analysis. Ann. Math. (2) **102**(1), 159–182 (1975). https://doi.org/10.2307/1970980
7. Beigi, S., Goodarzi, M.M.: Operator-valued Schatten spaces and quantum entropies. Lett. Math. Phys. **113**(5), Paper No. 91, 54 (2023). https://doi.org/10.1007/s11005-023-01712-9
8. Beigi, S., King, C.: Hypercontractivity and the logarithmic Sobolev inequality for the completely bounded norm. J. Math. Phys. **57**(1), 015206, 20 (2016). https://doi.org/10.1063/1.4934729
9. Blecher, D.P.: The standard dual of an operator space. Pac. J. Math. **153**(1), 15–30 (1992). http://projecteuclid.org/euclid.pjm/1102635970
10. Blecher, D.P.: Tensor products of operator spaces. II. Can. J. Math. **44**(1), 75–90 (1992). https://doi.org/10.4153/CJM-1992-004-5
11. Blecher, D.P., Gao, L., Xu, B.: Geometric influences on quantum boolean cubes (2024). https://arxiv.org/abs/2409.00224
12. Blecher, D.P., Le Merdy, C.: Operator Algebras and Their Modules—An Operator Space Approach. London Mathematical Society Monographs. New Series, vol. 30. The Clarendon Press Oxford University Press, Oxford (2004). http://dx.doi.org/10.1093/acprof:oso/9780198526599.001.0001
13. Blecher, D.P., Paulsen, V.I.: Tensor products of operator spaces. J. Funct. Anal. **99**(2), 262–292 (1991). http://dx.doi.org/10.1016/0022-1236(91)90042-4
14. Blecher, D.P., Smith, R.R.: The dual of the haagerup tensor product. J. Lond. Math. Soc. **45**(1), 126–144 (1992). https://doi.org/10.1112/jlms/s2-45.1.126
15. Bonami, A.: Étude des coefficients de Fourier des fonctions de $L^p(G)$. Ann. Inst. Fourier (Grenoble) **20**, 335–402 (1970). http://www.numdam.org/item?id=AIF_1970__20_2_335_0

© The Author(s), under exclusive license to Springer Nature Switzerland AG 2025

J. A. Chávez-Domínguez et al., *Operator Space Tensor Norms*, Lecture Notes in Mathematics 2379, https://doi.org/10.1007/978-3-031-96209-7

16. Carando, D., Galicer, D.: Natural symmetric tensor norms. J. Math. Anal. Appl. **387**(2), 568–581 (2012). https://doi.org/10.1016/j.jmaa.2011.09.027
17. Chávez-Domínguez, J.A.: Completely (q, p)-mixing maps. Ill. J. Math. **56**(4), 1169–1183 (2012). http://projecteuclid.org/euclid.ijm/1399395827
18. Chávez-Domínguez, J.A.: The Chevet-Saphar tensor norms for operator spaces. Houston J. Math. **42**(2), 577–596 (2016)
19. Chávez-Domínguez, J.A., Dimant, V., Galicer, D.: Operator p-compact mappings. J. Funct. Anal. **277**(8), 2865–2891 (2019). https://doi.org/10.1016/j.jfa.2019.03.001
20. Chávez-Domínguez, J.A., Dimant, V., Galicer, D.: Revisiting operator p-compact mappings. Rev. Mat. Complut. (2025). https://doi.org/10.1007/s13163-025-00527-7
21. Chevet, S.: Sur certains produits tensoriels topologiques d'espaces de Banach. Z. Wahrscheinlichkeitstheorie und Verw. Gebiete **11**, 120–138 (1969)
22. Choi, M.D.: Completely positive linear maps on complex matrices. Linear Algebra Appl. **10**, 285–290 (1975). https://doi.org/10.1016/0024-3795(75)90075-0
23. Defant, A., Floret, K.: Tensor Norms and Operator Ideals. North-Holland Mathematics Studies, vol. 176. North-Holland Publishing Co., Amsterdam (1993)
24. Defant, A., Wiesner, D.: Polynomials in operator space theory. J. Funct. Anal. **266**(9), 5493–5525 (2014). https://doi.org/10.1016/j.jfa.2014.02.018
25. Delgado, J.M., Piñeiro, C., Serrano, E.: Operators whose adjoints are quasi p-nuclear. Stud. Math. **197**(3), 291–304 (2010). https://doi.org/10.4064/sm197-3-6
26. Devetak, I., Junge, M., King, C., Ruskai, M.B.: Multiplicativity of completely bounded p-norms implies a new additivity result. Commun. Math. Phys. **266**(1), 37–63 (2006). https://doi.org/10.1007/s00220-006-0034-0
27. Diestel, J., Fourie, J.H., Swart, J.: The Metric Theory of Tensor Products: Grothendieck's résumé Revisited. American Mathematical Society, Providence (2008). https://doi.org/10.1090/mbk/052
28. Dimant, V., Fernández-Unzueta, M.: Biduals of tensor products in operator spaces. Stud. Math. **230**(2), 167–187 (2015). https://doi.org/10.4064/sm8292-1-2016
29. Dimant, V., Fernández-Unzueta, M.: Bilinear ideals in operator spaces. J. Math. Anal. Appl. **429**(1), 57–80 (2015). https://doi.org/10.1016/j.jmaa.2015.03.070
30. Effros, E.G., Kishimoto, A.: Module maps and Hochschild-Johnson cohomology. Indiana Univ. Math. J. **36**(2), 257–276 (1987). https://doi.org/10.1512/iumj.1987.36.36015
31. Effros, E.G., Ruan, Z.J.: On approximation properties for operator spaces. Int. J. Math. **1**(2), 163–187 (1990). https://doi.org/10.1142/S0129167X90000113
32. Effros, E.G., Ruan, Z.J.: Self-duality for the Haagerup tensor product and Hilbert space factorizations. J. Funct. Anal. **100**(2), 257–284 (1991). https://doi.org/10.1016/0022-1236(91)90111-H
33. Effros, E.G., Ruan, Z.J.: The Grothendieck-Pietsch and Dvoretzky-Rogers theorems for operator spaces. J. Funct. Anal. **122**(2), 428–450 (1994). http://dx.doi.org/10.1006/jfan.1994.1075
34. Effros, E.G., Ruan, Z.J.: Operator Spaces. London Mathematical Society Monographs. New Series, vol. 23. The Clarendon Press Oxford University Press, New York (2000)
35. Effros, E.G., Ruan, Z.J.: Operator space tensor products and Hopf convolution algebras. J. Oper. Theory **50**(1), 131–156 (2003)
36. Effros, E.G., Webster, C.: Operator analogues of locally convex spaces. In: Operator Algebras and Applications (Samos, 1996), NATO Advanced Study Institute Series C: Mathematical Physics Science, vol. 495, pp. 163–207. Kluwer Academic Publishers, Dordrecht (1997)
37. Effros, E.G., Junge, M., Ruan, Z.J.: Integral mappings and the principle of local reflexivity for noncommutative L^1-spaces. Ann. Math. (2) **151**(1), 59–92 (2000). http://dx.doi.org/10.2307/121112
38. Fawzi, O., Kochanowski, J., Rouzé, C., Van Himbeeck, T.: Additivity and chain rules for quantum entropies via multi-index schatten norms (2025). https://arxiv.org/abs/2502.01611
39. Gross, L.: Logarithmic Sobolev inequalities. Am. J. Math. **97**(4), 1061–1083 (1975). https://doi.org/10.2307/2373688

40. Grothendieck, A.: Résumé de la théorie métrique des produits tensoriels topologiques. Bol. Soc. Mat. São Paulo **8**, 1–79 (1953)
41. Haagerup, U.: The Grothendieck inequality for bilinear forms on C^*-algebras. Adv. Math. **56**(2), 93–116 (1985). doi:10.1016/0001-8708(85)90026-X, https://doi.org/10.1016/0001-8708(85)90026-X
42. Haagerup, U., Musat, M.: The Effros-Ruan conjecture for bilinear forms on C^*-algebras. Invent. Math. **174**(1), 139–163 (2008). https://doi.org/10.1007/s00222-008-0137-7
43. Ivanisvili, P., Nazarov, F.: On Weissler's conjecture on the Hamming cube I. Int. Math. Res. Not. IMRN (9), 6991–7020 (2022). https://doi.org/10.1093/imrn/rnaa363
44. Junge, M.: Factorization theory for spaces of operators. Habilitation Thesis. Kiel, 1996
45. Junge, M., Parcet, J.: Maurey's factorization theory for operator spaces. Math. Ann. **347**(2), 299–338 (2010). http://dx.doi.org/10.1007/s00208-009-0440-7
46. Junge, M., Nielsen, N.J., Ruan, Z.J., Xu, Q.: $\mathcal{COL}_p$ spaces—the local structure of non-commutative L_p spaces. Adv. Math. **187**(2), 257–319 (2004). http://dx.doi.org/10.1016/j.aim.2003.08.010
47. King, C.: Hypercontractivity for semigroups of unital qubit channels. Comm. Math. Phys. **328**(1), 285–301 (2014). https://doi.org/10.1007/s00220-014-1982-4
48. Kirchberg, E.: Exact C*-algebras, tensor products, and the classification of purely infinite algebras. In: Proceedings of the International Congress of Mathematicians, vol. 1, 2 (Zürich, 1994), pp. 943–954. Birkhäuser, Basel (1995)
49. Lindenstrauss, J., Pełczyński, A.: Absolutely summing operators in L_p-spaces and their applications. Stud. Math. **29**, 275–326 (1968). https://doi.org/10.4064/sm-29-3-275-326
50. Metger, T., Fawzi, O., Sutter, D., Renner, R.: Generalised entropy accumulation. Commun. Math. Phys. **405**(11), Paper No. 261, 43 (2024). https://doi.org/10.1007/s00220-024-05121-4
51. Montanaro, A., Osborne, T.J.: Quantum boolean functions. Chicago J. Theor. Comput. Sci. **1**, 1–45 (2010)
52. Oikhberg, T.: Completely bounded and ideal norms of multiplication operators and Schur multipliers. Integr. Equ. Oper. Theory **66**(3), 425–440 (2010). http://dx.doi.org/10.1007/s00020-010-1751-5
53. Oikhberg, T., Pisier, G.: The "maximal" tensor product of operator spaces. Proc. Edinb. Math. Soc. (2) **42**(2), 267–284 (1999). https://doi.org/10.1017/S0013091500020241
54. Paulsen, V.: Completely Bounded Maps and Operator Algebras. Cambridge Studies in Advanced Mathematics, vol. 78. Cambridge University Press, Cambridge (2002)
55. Persson, A.: On some properties of p-nuclear and p-integral operators. Stud. Math. **33**, 213–222 (1969)
56. Persson, A., Pietsch, A.: p-nukleare und p-integrale Abbildungen in Banachräumen. Stud. Math. **33**, 19–62 (1969)
57. Pietsch, A.: Absolut p-summierende Abbildungen in normierten Räumen. Studia Math. **28**, 333–353 (1966/1967)
58. Pietsch, A.: Operator ideals, Mathematische Monographien [Mathematical Monographs], vol. 16. VEB Deutscher Verlag der Wissenschaften, Berlin (1978)
59. Pietsch, A.: Operator ideals, North-Holland Mathematical Library, vol. 20. North-Holland Publishing Co., Amsterdam (1980). Translated from German by the author
60. Pisier, G.: Grothendieck's theorem for noncommutative C^*-algebras, with an appendix on Grothendieck's constants. J. Funct. Anal. **29**(3), 397–415 (1978). https://doi.org/10.1016/0022-1236(78)90038-1
61. Pisier, G.: Counterexamples to a conjecture of Grothendieck. Acta Math. **151**(3–4), 181–208 (1983). https://doi.org/10.1007/BF02393206
62. Pisier, G.: Factorization of linear operators and geometry of Banach spaces. CBMS Regional Conference Series in Mathematics, vol. 60. Published for the Conference Board of the Mathematical Sciences, Washington, DC. American Mathematical Society, Providence (1986). https://doi.org/10.1090/cbms/060
63. Pisier, G.: Exact operator spaces. Astérisque (232), 159–186 (1995). Recent advances in operator algebras (Orléans, 1992)

64. Pisier, G.: Non-commutative vector valued L_p-spaces and completely p-summing maps. Astérisque (247), vi+131 (1998)
65. Pisier, G.: Introduction to Operator Space Theory. London Mathematical Society Lecture Note Series, vol. 294. Cambridge University Press, Cambridge (2003)
66. Pisier, G.: Grothendieck's theorem, past and present. Bull. Am. Math. Soc. (N.S.) **49**(2), 237–323 (2012). https://doi.org/10.1090/S0273-0979-2011-01348-9
67. Pisier, G.: Tensor Products of C^*-algebras and Operator Spaces—the Connes-Kirchberg Problem. London Mathematical Society Student Texts, vol. 96. Cambridge University Press, Cambridge (2020). https://doi.org/10.1017/9781108782081
68. Pisier, G.: Tensor Products of C^*-algebras and Operator Spaces—the Connes-Kirchberg Problem. London Mathematical Society Student Texts, vol. 96. Cambridge University Press, Cambridge (2020). https://doi.org/10.1017/9781108782081
69. Pisier, G., Shlyakhtenko, D.: Grothendieck's theorem for operator spaces. Invent. Math. **150**(1), 185–217 (2002). https://doi.org/10.1007/s00222-002-0235-x
70. Reĭnov, O.I.: Approximation properties of order p and the existence of non-p-nuclear operators with p-nuclear second adjoints. Math. Nachr. **109**, 125–134 (1982). https://doi.org/10.1002/mana.19821090112
71. Ryan, R.A.: Introduction to Tensor Products of Banach Spaces. Springer Monographs in Mathematics. Springer-Verlag London Ltd., London (2002)
72. Saphar, P.: Applications p décomposantes et p absolument sommantes. Isr. J. Math. **11**, 164–179 (1972)
73. Watrous, J.: The Theory of Quantum Information. Cambridge University Press, Cambridge (2018)
74. Webster, C.J.: Local operator spaces and applications. Ph.D. thesis, University of California, 1997
75. Wiesner, D.: Polynomials in operator space theory. Ph.D. thesis, Carl von Ossietzky Universität Oldenburg, 2009
76. Wittstock, G., Betz, B., Fischer, H.J., Lambert, A., Louis, K., Neufang, M., Zimmermann, I.: What are operator spaces? https://www.math.uni-sb.de/ag/wittstock/OperatorSpace.pdf, Accessed Oct 19, 2019
77. Yew, K.L.: Completely p-summing maps on the operator Hilbert space OH. J. Funct. Anal. **255**(6), 1362–1402 (2008). http://dx.doi.org/10.1016/j.jfa.2008.06.019
78. Zalduendo, I.: Extending polynomials on Banach spaces—a survey. Rev. Un. Mat. Argentina **46**(2), 45–72 (2005/2006)

List of Symbols

$(\mathfrak{A}, \mathbf{A}) \sim \alpha$; mapping ideal associated to an o.s. tensor norm, 110

$(\mathfrak{A}, \mathbf{A})$; normed mapping ideal, 93

$\mathfrak{A}^{\mathrm{dual}}$; dual of the mapping ideal $\mathfrak{A}$, 105

$\mathcal{A}$; completely approximable mappings, 94

$\mathfrak{A}^*$; adjoint of the mapping ideal $\mathfrak{A}$, 120

$\mathfrak{A}^{\mathrm{inj}}$; injective hull of the mapping ideal $\mathfrak{A}$, 100

$\mathfrak{A}^{\mathrm{max}}$; maximal hull of the mapping ideal $\mathfrak{A}$, 108

$\mathfrak{A}^{\mathrm{min}}$; minimal kernel of the mapping ideal $\mathfrak{A}$, 121

$\mathfrak{A}^{\mathrm{sur}}$; surjective hull of the mapping ideal $\mathfrak{A}$, 103

$\mathfrak{A}_\alpha$; α-continuous mapping ideal, 99

$\alpha\backslash$; completely right-injective hull of α, 133

$\alpha/$; completely right-projective hull of α, 138

α'; dual norm of α, 60

α^*; adjoint norm of α, 60

α^t; transpose norm of α, 20

$\alpha^{\rightarrow}$; right-finite hull of α, 137

$\backslash\alpha/$; completely projective hull of α, 142

$\backslash\alpha$; completely left-projective hull of α, 142

$\overleftarrow{\alpha}$; cofinite hull of α, 39

$\overrightarrow{\alpha}$; finite hull of α, 39

$/\alpha$; completely left-injective hull of α, 136

$/\alpha\backslash$; completely injective hull of α, 136

$\mathbf{A}^{\mathrm{dual}}$; dual norm, 105

$\mathbf{A}^*$; adjoint ideal norm, 120

$\mathbf{A}^{\mathrm{inj}}$; injective hull norm, 101

$\mathbf{A}^{\mathrm{max}}$; maximal ideal norm, 108

$\mathbf{A}^{\mathrm{min}}$; minimal ideal norm, 122

$\mathbf{A}^{\mathrm{sur}}$; surjective hull norm, 103

$\mathcal{B}(H)$; space of bounded operators on the Hilbert space H, 2

J. A. Chávez-Domínguez et al., *Operator Space Tensor Norms*, Lecture Notes in Mathematics 2379, https://doi.org/10.1007/978-3-031-96209-7

C; Column space, 10

CB; completely bounded mappings, 2

$\|\cdot\|_{\mathrm{cb}}$; completely bounded norm, 2

χ_S; Walsh function with support in the set S, 74

CC; completely compact mappings, 99

$c_0(\{E_\gamma : \gamma \in \Gamma\})$; c_0-sum of operator spaces, 4

$E\widehat{\otimes}_\alpha F$; completion of the tensor product $E \otimes_\alpha F$, 20

Δ_p; operator space structure on $S_p(H) \otimes E$ induced from $S_p[H; E]$, 69

d_p^B; right p-Chevet-Saphar (Banach space) tensor norm, 75

d_p; right p-Chevet-Saphar tensor norm, 24

eh; extended Haagerup tensor norm, 37

ι^{ex}; exactly integral norm, 95

$\mathcal{E}(\lambda)$; extended λ-o.s. tensor norms, 53

$\mathcal{I}^{\mathrm{ex}}$; exactly integral mappings, 95

$\overline{E}$; completion of E, 5

$\mathcal{F}$; finite rank mappings, 94

Γ_c; mappings factoring through column spaces, 97

Γ_r; mappings factoring through row spaces, 97

γ_2^c; mappings factoring through column spaces norm, 98

γ_2^r; mappings factoring through row spaces norm, 98

g_p; left p-Chevet-Saphar tensor norm, 24

H_c; Column Hilbert space, 11

H_r; Row Hilbert space, 11

h; Haagerup tensor norm, 8

$\alpha \cap \beta$; intersection of two o.s. tensor norm, 36

ι; completely integral norm, 95

$\mathcal{I}$; completely integral mappings, 95

JCB; jointly completely bounded bilinear mappings, 3

$\|\cdot\|_{\mathrm{jcb}}$; jointly completely bounded norm, 3

$K_\infty(E) = S_\infty[E]$; E-valued ∞-Schatten space, 85

κ_E; the canonical injection into the bidual, 3

κ_p; operator p-compact norm, 96

$\mathcal{K}$; compact operators on ℓ_2, 10

$\mathcal{K}_p$; operator p-compact mappings, 96

$E_1 \oplus_1 E_2$; ℓ_1-sum of two operator spaces, 5

$E_1 \oplus_\infty E_2$; ℓ_∞-sum of two operator spaces, 4

$L_p(\mathrm{tr}_n)$; noncommutative L_p space associated to the normalized trace tr_n, 76

$\ell_1(\{E_\gamma : \gamma \in \Gamma\})$; ℓ_1-sum of operator spaces, 5

ℓ_2^k; k-dimensional Hilbert space, 1

$\ell_\infty(\{E_\gamma : \gamma \in \Gamma\})$; ℓ_∞-sum of operator spaces, 4

λ; λ-o.s. tensor norm, 32

$M_{n,m}(E)$; set of $n \times m$-matrices of elements in E, 1

$M_n(E)$; set of $n \times n$-matrices of elements in E, 1

MB; multiplicatively bounded bilinear mappings, 3

Max(W); maximal operator space structure of the Banach space W, 4

Index

J. A. Chávez-Domínguez et al., *Operator Space Tensor Norms*, Lecture Notes
in Mathematics 2379, https://doi.org/10.1007/978-3-031-96209-7

LECTURE NOTES IN MATHEMATICS   Springer

Editors in Chief: J.-M. Morel, B. Teissier;

Editorial Policy

1. Lecture Notes aim to report new developments in all areas of mathematics and their applications – quickly, informally and at a high level. Mathematical texts analysing new developments in modelling and numerical simulation are welcome.

   Manuscripts should be reasonably self-contained and rounded off. Thus they may, and often will, present not only results of the author but also related work by other people. They may be based on specialised lecture courses. Furthermore, the manuscripts should provide sufficient motivation, examples and applications. This clearly distinguishes Lecture Notes from journal articles or technical reports which normally are very concise. Articles intended for a journal but too long to be accepted by most journals, usually do not have this "lecture notes" character. For similar reasons it is unusual for doctoral theses to be accepted for the Lecture Notes series, though habilitation theses may be appropriate.

2. Besides monographs, multi-author manuscripts resulting from SUMMER SCHOOLS or similar INTENSIVE COURSES are welcome, provided their objective was held to present an active mathematical topic to an audience at the beginning or intermediate graduate level (a list of participants should be provided).

   The resulting manuscript should not be just a collection of course notes, but should require advance planning and coordination among the main lecturers. The subject matter should dictate the structure of the book. This structure should be motivated and explained in a scientific introduction, and the notation, references, index and formulation of results should be, if possible, unified by the editors. Each contribution should have an abstract and an introduction referring to the other contributions. In other words, more preparatory work must go into a multi-authored volume than simply assembling a disparate collection of papers, communicated at the event.

3. Manuscripts should be submitted either online at www.editorialmanager.com/lnm to Springer's mathematics editorial in Heidelberg, or electronically to one of the series editors. Authors should be aware that incomplete or insufficiently close-to-final manuscripts almost always result in longer refereeing times and nevertheless unclear referees' recommendations, making further refereeing of a final draft necessary. The strict minimum amount of material that will be considered should include a detailed outline describing the planned contents of each chapter, a bibliography and several sample chapters. Parallel submission of a manuscript to another publisher while under consideration for LNM is not acceptable and can lead to rejection.

4. In general, **monographs** will be sent out to at least 2 external referees for evaluation.

   A final decision to publish can be made only on the basis of the complete manuscript, however a refereeing process leading to a preliminary decision can be based on a pre-final or incomplete manuscript.

   Volume Editors of **multi-author works** are expected to arrange for the refereeing, to the usual scientific standards, of the individual contributions. If the resulting reports can be forwarded to the LNM Editorial Board, this is very helpful. If no reports are forwarded or if other questions remain unclear in respect of homogeneity etc, the series editors may wish to consult external referees for an overall evaluation of the volume.

5. Manuscripts should in general be submitted in English. Final manuscripts should contain at least 100 pages of mathematical text and should always include

   – a table of contents;
   – an informative introduction, with adequate motivation and perhaps some historical remarks: it should be accessible to a reader not intimately familiar with the topic treated;
   – a subject index: as a rule this is genuinely helpful for the reader.
   – For evaluation purposes, manuscripts should be submitted as pdf files.

6. Careful preparation of the manuscripts will help keep production time short besides ensuring satisfactory appearance of the finished book in print and online. After acceptance of the manuscript authors will be asked to prepare the final LaTeX source files (see LaTeX templates online: https://www.springer.com/gb/authors-editors/book-authors-editors/manuscriptpreparation/5636) plus the corresponding pdf- or zipped ps-file. The LaTeX source files are essential for producing the full-text online version of the book, see http://link.springer.com/bookseries/304 for the existing online volumes of LNM). The technical production of a Lecture Notes volume takes approximately 12 weeks. Additional instructions, if necessary, are available on request from lnm@springer.com.

7. Authors receive a total of 30 free copies of their volume and free access to their book on SpringerLink, but no royalties. They are entitled to a discount of 33.3 % on the price of Springer books purchased for their personal use, if ordering directly from Springer.

8. Commitment to publish is made by a *Publishing Agreement*; contributing authors of multiauthor books are requested to sign a *Consent to Publish form*. Springer-Verlag registers the copyright for each volume. Authors are free to reuse material contained in their LNM volumes in later publications: a brief written (or e-mail) request for formal permission is sufficient.

Addresses:
Professor Jean-Michel Morel, CMLA, École Normale Supérieure de Cachan, France
E-mail: moreljeanmichel@gmail.com

Professor Bernard Teissier, Equipe Géométrie et Dynamique,
Institut de Mathématiques de Jussieu – Paris Rive Gauche, Paris, France
E-mail: bernard.teissier@imj-prg.fr

Springer: Ute McCrory, Mathematics, Heidelberg, Germany,
E-mail: lnm@springer.com